LIFE ITSELF

Complexity in Ecological Systems Series

COMPLEXITY IN ECOLOGICAL SYSTEMS SERIES

T.F.H. Allen and David W. Roberts, Editors

Robert V. O'Neill, Adviser

LIFE ITSELF

A Comprehensive Inquiry Into the
Nature, Origin, and Fabrication of Life

———

ROBERT ROSEN

COLUMBIA UNIVERSITY PRESS · NEW YORK

COLUMBIA UNIVERSITY PRESS

NEW YORK, CHICHESTER, WEST SUSSEX

Publishers Since 1893

Copyright © 1991 Columbia University Press

All rights reserved

Library of Congress Cataloging-in-Publication Data

Rosen, Robert, 1934–
    Life itself : a comprehensive inquiry into the nature, origin, and fabrication of life /
Robert Rosen.
        p.   cm. — (Complexity in ecological systems series)
    Includes index.
    ISBN 978-0-231-07564-0 (cloth : alk. paper) — 978-0-231-07565-7 (pbk. : alk. paper)
    1. Life (Biology) 2. Life—Origin. 3. Biological systems. 4. Biology—Philosophy.
I. Title. II. Series.
QH325.R57   1991
574—dc20

91-3110
CIP

♾

Casebound editions of Columbia University Press books are Smyth-sewn and printed on
permanent and durable acid-free paper.

To the memory of
Nicolas Rashevsky
and
James F. Danielli,
who would have been interested

Contents

To the memory of
Nicolas Rashevsky
and
James F. Danielli,
who would have been interested

Foreword

T.F.H. Allen and David W. Roberts

TWO DEMANDS are being made upon the community of ecologists: that their discipline be increasingly in a predictive mode; and that ecologists be prepared as never before to move up-scale and consider large-scale systems. Leaps in the technology for data acquisition and processing make it feasible to deal with large-scale phenomena in relatively fine grain terms. These imperatives require, and these opportunities allow, ecologists to deal with complex systems. All ecologists are aware that there is much complexity in almost all ecological systems. So impressive is ecological complexity that one becomes convinced that there is something in the very nature of ecological material that is complex. However, it emerges that nature itself is neither complex nor simple. Complexity is a matter of how the observer specifies the system either explicitly or implicitly in the way questions are cast. What makes ecology complex is the challenge of the questions we dare to ask of nature. When the whole entity displaying the phenomenon is scaled much larger than the entities used as explanatory principles, then the system is complex. It is therefore the urgency of certain questions that presses ecologists into the realm of complex systems.

In complex systems, abdication of the scientists' responsibility for specification and scaling leads to confusion and contradiction. Complexity requires a strict and consistent epistemology. The paradigm of this series is that complexity is tractable but demands parallel description at many explicitly specified levels. In the face of complexity it is essential to distinguish model and observables from the material system, and to recognize that the model must invoke a scale and a point of view. With that in mind, the books in this series explore many facets of ecology broadly defined.

The first two books in the series, *Life Itself* by Robert Rosen and

Toward a Unified Ecology by T. F. H. Allen and T. W. Hoekstra, approach theory in the most general terms. The present volume is as much about the general properties of life as it is about ecological systems narrowly defined. Such is the scope of this project that a book about the generalities of living systems is well within the series' purview. By contrast, in preparation are several books that take specific facets of the unified ecology and lay out their complexities. Some volumes in the series will be on distinctly applied topics like economic ecology or the larger issues in environmental regulation. A majority of the volumes will, however, deal with ideas in basic ecology, such as complexity in spatially defined landscape systems and the difficulties of aggregation as one moves between ecological hierarchical levels.

While the theory will employ whatever mathematics is necessary, the editors hope for a style of theory that is understandable because of its intuitive appeal and clear relevance. The series is intended to offer a useful context for specific research projects on particular types of organisms, perhaps in some local setting. The main body and mainstream of ecological research will always have that quality of real organisms or tangible processes in real places. In hierarchical systems, the upper level gives meaning to the level of focus, the lower level. There need be only a small difference between ecological research that is prosaic and isolated as opposed to research in those same areas that deservedly commands attention from pertinent related areas of research, and the ecological research community at large. The editors will try to have the series offer a means of casting mainstream research in a meaningful context. By presenting a milieu in which the mainstream of ecology moves forward, we expect to enable a more coherent ecology, but one performed by specialists within the various subdisciplines as appropriate.

Preface

FOR WHOM is this book intended? I do not know. The book itself has no pragmatic purpose of which I am aware. It is thus for anyone who wants to claim it. My sentiments in this regard are exactly expressed in the words with which Johann Sebastian Bach dedicated the third part of his *Clavierübung:*

> *Bestehend in verschiedenen Vorspielen . . .*
> *Denen Liebhabern und besonders denen*
> *Kennern von vergleichen Arbeit*
> *zur Gemuths Ergezung.*

> (Consisting of sundry preludes . . . for those
> who love such things, and particularly for those
> who are familiar with comparable work, for the
> delight of their souls.)

This book represents part of the outcome and present status of about thirty years' work on the problem "What is life?" Some of it has appeared elsewhere, in the form of journal articles, or in some of my other books. But most of it, especially the epistemological considerations at the heart of it, has accumulated unpublished and unpublishable, except in this form.

I must add that writing this text is the hardest thing I have ever tried to do, much harder than doing the research it embodies. The problem was to compress a host of interlocking ideas, drawn from many sources, which coexist happily in my head, into a form coherently expressible in a linear script. Moreover, there was the problem of trying to indicate the richness of many of the ideas, which in themselves want to ramify off in many directions, while keeping them focused on the primary problem. This last has involved a rigorous, and at times exquisitely painful, selection process;

I could easily have expanded almost every chapter, and indeed many of the chapter sections, into separate volumes each the size of the present one.

I have tried to address these problems of organization and exposition in the same way as I approach research, namely, to try to let the problems tell me what to do. Everything to be found herein, and the manner in which it is expressed, is what it is because it seemed to me to be necessary. And about each, I can say in truth that it is the best I can do. It is for the reader to say whether this has been enough.

Note to the Reader

THIS BOOK, like most books, has a primary structure and a tertiary structure. The primary structure is, of course, the linear string of symbols, words, paragraphs, and chapters that is mandated by the nature of the language. But just as with a molecule of globular protein, which must fold in order to become active, this linear string must be folded back on itself, so as to bring parts remote on the string into close contact. This particular volume is very heavily folded indeed. I have tried to indicate at least some of the major folds via an elaborate system of cross-references, sometimes back to parts of the string we have already traversed, and sometimes ahead to parts yet to come. Many other folds are not so indicated, but I hope that enough of them have been specified so that the remainder are obvious.

In any case, an overview of the fully folded structure, and a word about why it is folded as it is, may be helpful at the outset.

First and foremost, it should always be borne in mind that this book is about biology. More specifically, it is a report of where the question "What is life?" has taken me in the quest for an answer. I am well aware that most of the ideas developed herein seem, in isolation, to have little to do with conventional biology. But ideas fold too; even those that seem most remote in terms of their initial origin, and even their content, may turn out to lie very close together indeed in some appropriate topology.

Indeed, at the risk of pushing the folding metaphor too far, this entire volume is an attempt to show how the ideas developed herein constitute, in effect, a single enzyme geared to lyse the problem "What is life?"

It has turned out that, in order to be in a position to say what life *is*, we must spend a great deal of time in understanding what life is *not*. Thus, I will be spending a great deal of time with mechanisms and machines, ultimately to reject them, and replace them with something else. This is in fact the most radical step I shall take, because for the past three centuries, ideas of mechanism and machine have constituted the very essence of the adjective "scientific"; a rejection of them thus seems like a rejection of science itself.

But this turns out to be only a prejudice, and like all prejudices, it has disastrous consequences. In the present case, it makes the question "What is life?" unanswerable; the initial presupposition that we are dealing with mechanism already excludes most of what we need to arrive at an answer. No amount of refinement or subtlety within the world of mechanism can avail; once we are in that world, what we need is already gone. Thus, we must retreat to an earlier epistemological stage, before the assumptions that characterize mechanisms have been made.

The stage to which we retreat is embodied in what we call Natural Law, which seems to me to be the bare minimum required to do science at all. The essence of it lies in what we call the modeling relation. Roughly speaking, this involves only the bringing of two systems of entailment into congruence. Modeling relations provide the thread that ties everything in this volume together.

From this perspective, the hypotheses of mechanism turn out to be only a very special way of embodying Natural Law, and correspondingly, a very limited way. As it turns out, mechanism presupposes the identity of two modes of modeling that can be very different from each other and that we exhibit under a wide variety of rubrics: analytic/synthetic, syntactic/semantic, product/coproduct, etc. Presupposition of their identity thus creates a degeneracy, a nongenericity; it is precisely this very degeneracy that has become identified with science itself. But it is, from my perspective, an impediment; when I remove it, I glimpse a whole new world of possibilities, with exactly the same claim to be called science as mechanism does. Only in such a world, as I argue, do we find the resources to grapple with the question "What is life?"

My arguments will be in no way speculative. At each stage, my conclusions are forced by the nature of the problem itself, and to where I have been led by preceding stages. This procedure in fact leaves no room for speculation at all. In my view, theory is the very antithesis of speculation, despite an all-pervading confusion between the two, an inability to tell a hypothesis from a conclusion.

I shall spend a great deal of time discussing the concept of state and recursive state transitions. These represent the central features of mechanism. Recursiveness is the key property; it connotes a situation in which entailment can in some sense be moved from domain to range. Only very special entailment structures allow this; accordingly, the presuppositions of mechanisms automatically mandate a corresponding impoverishment of entailment. In another language, there is not much causality in a mechanism; almost everything about it is unentailed.

As I shall argue at the end, organisms sit at the other extreme of the entailment specturm than mechanisms do; almost everything about them is entailed by something else about them. This is why an initial presupposition of mechanism is so devastating; it restricts us to fragments, pieces that individually can be regarded as mechanisms all right but that cannot be articulated or combined within those confines.

There are two ways out of the world of mechanism that I shall describe. One of them concerns the circle of ideas called *relational,* which I develop at some length from various points of view. These ideas are concerned with function; because this word "function" has no general mechanistic counterpart, it is accordingly regarded as unscientific or prescientific. Nevertheless, there is a class of mechanisms (called *machines*) that do admit the concept of function and that accordingly have relational descriptions. As I show, already in machines, the relational description involves a divorcing or separation of function (embodied in the relational description) from structure (embodied in the mechanistic ones). Relational descriptions retain meaning even when structural ones do not. Hence relational ideas can provide one kind of exit from the cage of mechanism.

Another exit arises from the application of a familiar formal procedure, the taking of limits. In general, the limit of a sequence of mechanisms need not be a mechanism; the limit of a sequence of mechanical models of a system may still be a model but not a mechanical one. A system of this kind will be called a *complex* system (although this usage of the term "complex" differs somewhat from that of most other authors).

Thus, I come at the end to propose an answer to the question "What is life?" In a nutshell the answer can be expressed this way: an organism is a material system that (1) is complex, and (2) admits a certain kind of relational description.

The consequences of these ideas are indeed radical. In a sense, physics shrinks and biology expands. Physics as we know it today is, almost entirely, the science of mechanism, and mechanisms, as I argue, are very special as material systems. Biology involves a class of systems more general than mechanisms. In fact the relative positions of physics and biology become interchanged; rather than physics being the general and biology the special, it becomes more the other way around. There are other consequences as well, to which I draw attention in the text as I come to them.

This volume concludes, as it must, with a discussion of the origin problem, which I pose in the context of *fabrication.* In complex systems, as

I argue, the chain of entailments involved in fabrication of a system is completely different from that underlying the operation of the system. In mechanisms, on the other hand, owing partly to the causal impoverishment to which I have already alluded, the same kinds of causal chains underlie both. On the other hand, an entirely new feature manifests itself precisely here; because organisms embody so much entailment, a relational theory of organisms is also a general theory of fabrication. These remarks will provide the point of departure for subsequent volumes.

On the basis of Natural Law, then, we are allowed to draw quite a number of drastic conclusions. What Natural Law gives us are certain categories of formal objects, the category of all models of a given material system. Science itself is bound up essentially with the structures of such categories, and the relations that can exist among them (which in turn reflect themselves as relations of metaphor, analogy, etc., among the natural systems represented by the models). Ideas drawn from the Theory of Categories correspondingly permeate my enterprise; it provides the natural language for expressing both my arguments and my conclusions.

This book necessarily involves a lot of formalism, often as subject, often as tool, sometimes as both. But it is not formal in the usual sense; what I am talking about does not admit formal treatment in that sense. My subject matter is rather the role of formalisms in describing entailments in the natural world and what the formalisms *mean* in such a context. Mathematicians, both pure and applied, often find such considerations repugnant, but that is too bad; it cannot be helped.

I conclude these preliminaries with a personal word, regarding the genesis of these ideas reported below. I started from early childhood with a lust to do biology, which I retain. In pursuit of this lust, I acquired a great deal of mathematics; to me, this seemed natural, because as was said before, ideas fold. My main preoccupation in those days was to learn all about operator algebras, the language of quantum mechanics, which I then believed would be sufficient. Almost by accident, I also absorbed a lot of Theory of Categories. It has turned out that the latter was much more important than the former.

My first steps in actually trying to do biology happened to be relational. I was early convinced that such considerations provided legitimate models, descriptions of natural systems that had as much right to be called models as any system of differential equations.

The trouble came when I tried to integrate relational and structural descriptions of the same biological systems. They did not seem to want to go together gracefully. Yet they *must* go together, being alternate descriptions of the same systems, the same material reality. Moreover, I needed both; biology seemed to require both.

They say that all science must start from experience. Mine was that relational models and mechanical models, drawn from physical analysis through reductionism, were not going together. That was a fact. My conclusion from that fact was that I was simply being stupid, or else there were some deep and essential things embodied in that fact. I was never able to rule out the first possibility, but the possibility of the second is what has led circuitously, in the course of time, to what is chronicled herein. In short, this is where I have been led by merely following the problem. It is the problem that imbues the path itself with whatever intrinsic logic is discernable in retrospect.

And so, I repeat, what is in this book is about biology. It directly addresses the basic question of biology: "What is life?" And nothing can be more biological than that.

LIFE ITSELF

PRAELUDIUM

SOME WEEKS after completing the manuscript of this monograph, I had to prepare the essay that follows. It constitutes the text of a talk I was asked to give at a scientific conference, and the title was predicated for me in advance. Naturally, like all preludes, it was written in the light of what had preceded it. I believe that, as it turned out, it also constitutes the best possible introduction to the rest of the volume. It is not an abstract or summary in the usual sense; rather, its relation to the text itself is (to use another currently popular image) that of a piece of a fractal to the whole thing.

I believe that by reading this introductory essay, this praeludium, the reader will obtain a clearer idea of the whys and wherefores of what follows than any summary could provide.

"Hard" Science and "Soft" Science

Some years ago, the novelist C. P. Snow drew attention to a dualism that permeates and poisons the intellectual life of our times, a dualism between science and art, between science and humanism.

The dualism to which Snow, among others, drew attention is indeed real. It has always been real and has existed since human beings first learned to think and to communicate their thoughts. But the situation is, and always has been, far worse than Snow has depicted. He painted a picture of science itself as a kind of pure phase, and its relation to other aspects of our culture as a kind of phase separation; scientists and humanists separating from each other as oil separates from water, through a preference of like for like, and an antipathy of like for unlike. But the dualities that Snow depicted also permeate science itself.

I have, much against my will, been immersed my whole life in one of these dualities, namely, the antagonism between "theory" and "experiment." My subject matter herein is another, in fact closely related duality, that between "hard" science and "soft" science, between quantitative and qualitative, between "exact" and "inexact."

This duality is not to be removed by any kind of tactical accommodation, by any superficial effort of conciliation or ecumenicism. The antipathies generated by the duality itself are only symptoms of a far deeper situation, which has roots partly in specific subject matter, partly in individual aspirations, and most important, in the embracing of mutually incompatible *weltanschauungen*, which reflect the deepest aspects of temperament and personality. It is thus not a matter of logical argumentation or persuasion that is involved here; it is a matter more akin to religious conversion.

In what follows, I discuss the duality between qualitative and quantitative. As we will see, in the sciences this dichotomy rests on (generally unrecognized) presuppositions about the nature of material reality and on how we obtain knowledge about it. I will then show that these presuppositions themselves have formal, mathematical counterparts, which allow us to reflect this scientific dualism into an exactly parallel one that exists within mathematics itself. This mathematical form of the dualism is centered around the notion of *formalization;* it can be expressed as the duality between syntactics and semantics; between what is true by virtue of form alone, independent of any external referents, and what is not.

The virtue of doing this is that there is a theorem (Gödel's Theorem) that actually resolves the issue, at least in part. When we pull this theorem back into a scientific context, by looking at its epistemological correlates, we obtain thereby some new and deep insights into the duality between quantitative and qualitative; between "hard" and "soft." I think that all concerned will find some surprises in this exercise.

Naturally, in this brief space, I can give only the most cursory sketch of the ideas involved. But I hope that enough will be said to provoke some reappraisals on both sides of the duality.

QUALITIES AND QUANTITIES IN THE SCIENCES

I can perhaps best illustrate the dichotomy between quality and quantity in the sciences with two quotations. The first is due to Ernest Rutherford:

> Qualitative is *nothing but* poor quantitative.

The emphasis is mine. The second quotation is due to Robert Hutchins, a man no less clever than Rutherford:

A social scientist is a person who counts telephone poles.

I chose these, out of the countless others that could be used, because these words are fighting words; they most vividly exhibit the emotional character of the issues involved.

Obviously, for Rutherford, *everything* we call a quality or a percept is expressible in terms of numerical magnitudes, without loss or distortion. For Rutherford, therefore, every quality can be quantitated, and hence, measured and/or computed. For Rutherford, science does not begin until this quantitation is made, until crude and inexact talk about quality is replaced by precise, exact, and completely equivalent talk about numbers. Indeed, to discuss qualities in any other terms is contemptible ("poor quantitative").

Hutchins tacitly accepts Rutherford's equation of "science" with "quantitative," but for him, this makes the phrase "social scientist" an oxymoron, a contradiction in terms. For Hutchins the features or qualities of a social structure that are of interest or importance are precisely those that are *unquantifiable;* conversely, anything we *can* count is trivial or irrelevant ("telephone poles").

Could anything more clearly exhibit the issues involved here? Let me articulate a few of them. Can arbitrary qualities (the stuff of perception) be equivalently expressed in terms of a certain limited subset of elementary qualities (those we can measure numerically)? Of so, how? If not, what does it mean to have a science of such qualities? What relationships can exist between such sciences (if indeed, any at all)? It is clear that these issues and others like them involve the deepest aspects of the relation between the perceiving mind and the perceptual universe, that the attitudes expressed by Rutherford and Hutchins in the quotations above involve radically different views of these questions, and that they cannot both be right.

Rutherford's view, that every perceptual quality can, and must, be expressible in numerical terms, is associated with the viewpoint commonly called *reductionism.* In practice, reductionism actually asserts much, much more than this; in its most extreme form, it actually identifies a specific family of elementary numerical qualities (and the procedures for measuring them, at least in principle) and anchors them in *physics*. How this is done is a very long story, which I cannot enter upon here but which can be found in an earlier work. [1] According to this view, there *is* no other science than

1. R. Rosen, *Anticipatory Systems* (New York: Pergamon Press, 1985). Hereafter cited in text as *AS*.

physics; everything else we call a science is ultimately a special case of physics.

Hutchins' view, on the other hand, is that physics, in this sense, is itself only a very special science, limited precisely to those very special qualities that happen to also be quantifiable. Accordingly, reductionism is wrong in principle; each science must have its own character and its own procedures, shaped by the specific class of qualities with which it must deal. To Hutchins, who was throughout his life concerned with qualities of social systems that pertain (in the broadcast sense) to politics, it was *precisely* to the extent that something was quantifiable, or expressible in terms of numerical magnitudes, that it was irrelevant. Thus, in effect, Hutchins is inverting Rutherford's dictum; he is asserting that *quantitative is poor qualitative.*

There is nothing inherently illogical, or even unscientific, about either of these positions. They differ so radically because they clearly start from entirely different philosophical presuppositions about the nature of the perceptual world and the relation of the perceiver to the percepts. It is clear that Hutchins and Rutherford could hardly communicate beyond superficialities; they could not be friends. That is precisely the dualism between "hard science," personified here by Rutherford, and "soft science," personified by Hutchins.

There are ways out of this impasse, but they are not palatable to either party. They involve a recognition that mathematics has more to offer besides numbers and a corresponding recognition that perceptual qualities may be expressed in terms of them (i.e., that "measurement" is not the simple thing Rutherford thought it was). But once again, the exploration of these matters is not my present purpose.

Syntax and Semantics

I am now going to do something that would bother both Rutherford and Hutchins, though in different ways. I am going to illuminate the duality they personify by looking at a cognate situation in an entirely different realm, the realm of mathematics. I do this for two separate reasons: first, because it *is* illuminating, and second, because the *fact* that it is illuminating is one that neither Hutchins nor Rutherford could account for.

Let me begin with a few words about the relations existing between the mathematical universe and the perceptual one. It is a fact of experience, for instance, that

$$2 \text{ sticks} + 3 \text{ sticks} = 5 \text{ sticks.}$$

On its face, this is a proposition about *sticks*. But it is not the same kind of proposition as, say, "sticks burn" or "stocks float." It differs from them in that it is also about something else besides sticks, and that "something else" takes us into the world of mathematics.

The mathematical world is *embodied* in percepts but exists independent of them. "Truth" in the mathematical world is likewise manifested in, but independent of, any material embodiment and is thus outside of conventional perceptual categories like space and time. These facts have indeed, from the time of Pythagoras on, spawned another profound dualism, a dualism between idealism (which at root is an attempt to extend the reality of number to the rest of the perceptual universe) and materialism (which is an attempt to include "mathematical reality" inside conventional perceptual realms).

But of this I need not speak. To motivate our discussion, it is enough to observe that both science, the study of phenomena, and mathematics are in their different ways concerned with systems of *entailment*, causal entailment in the phenomenal world, inferential entailment in the mathematical. At root, where Hutchins and Rutherford differ is precisely in their views about entailment, about what is entailed from a datum and about how that datum is itself entailed. Hence, at a sufficiently deep level, the controversy between them, and the dualism they represent, pertains to entailment itself, entailment in the abstract, free of any qualifying adjectives like "causal" or "inferential."

It is in this sense that I turn to the mathematical world in order to illuminate what it tells us about entailment. That is, I will be talking about entailment, rather than about mathematics, just as, in the example above, I could talk about number while apparently talking about sticks.

Mathematics over the past century has given little evidence that it is concerned with eternal, timeless, and hence, unarguable truth. On the contrary, contemporary mathematics is filled with (no pun intended) chaos and turbulence, bespeaking a profound internal instability. Historically, we are still witnessing what (it is hoped) are the transients arising from two profound shocks: the overthrow of Euclid and the discovery of inconsistencies (paradoxes) in Set Theory.

To be sure, most practicing mathematicians, like most practical (empirical) scientists, go on about their business, indifferent to such matters, convinced to the depths of their soul about the reliability of what they do; like corks, they believe they will float on calm and troubled waters alike. But I am speaking of foundations, and it is here we shall look.

The two great shocks of which I spoke above have coalesced, beginning in the early years of the present century, into a frantic concern with *consistency*, with a demand that a system of inferential entailments (e.g., a set of axioms or production rules, operating on a set of given propositions or postulates) be free of internal or logical contradictions. How can we be sure that a system of entailment, e.g., a mathematical system in the broadest sense, is consistent? I shall be concerned with one particular kind of answer given to this question, an answer championed by David Hilbert, which can be summed up in a single word: *formalization*.

Hilbert and others thought they had traced down the ultimate source of all the difficulties in mathematics. They pointed out that propositions in mathematics are nominally *about* something; i.e., they have meanings that involve referents outside themselves. Thus, for instance, in Euclid, the word "triangle" is not *just* an array of letters to be manipulated in a certain way; it *refers* to a rich and vivid kind of geometric object. And even beyond that, it even refers to things in the phenomenal world. In that sense, any Euclidean proposition containing the word "triangle" can be thought of as describing a *percept* or *quality* manifested by this external referent.

Thus, according to this analysis, mathematical truth had come to involve two distinct aspects, one pertaining to how we are allowed to manipulate the word "triangle" from one proposition to another, and another pertaining to the actual referents of that word. I will call the former *syntactic* truth, the latter, *semantic* truth. Hilbert and his colleagues argued that it was precisely by allowing semantic truth into mathematics at all (i.e., in the admissibility of regarding a mathematical proposition as the description of a percept or quality of something, allowing a mathematical proposition to *refer* to something) that all the difficulty arises.

Hilbert and his formalistic school actually asserted much more than this. They argued that what we have called *semantic truth could always be effectively replaced by more syntactic rules*. In other words, any external referent, and any quality thereof, could be pulled inside a purely syntactic system. By a purely syntactic system, they understood: (1) a finite set of *meaningless* symbols, an alphabet; (2) a finite set of rules for combining these symbols into strings or formulas; (3) a finite set of production rules for turning given formulas into new ones. In such a purely syntactic system, *consistency is guaranteed.*

The best statement I have even seen of the formalistic program is that given by Kleene; it does no harm to quote it again:

> (Formalization) will not be finished until all the properties or undefined terms which matter for the deduction of theorems have been expressed by axioms. Then it should be possible to perform the deductions treating the technical terms as words in themselves without meaning. For to say that they have

meanings necessary to the deduction of the theorems, other than what they derive from the axioms which govern them, amount to saying that not all of their properties which matter for the deductions have been expressed by axioms. When the meanings of the technical terms are thus left out of account, we have arrived at the standpoint of formal axiomatics.[2]

The idea that all truth can be expressed as pure syntactic truth, which is the essence of the formalist position in mathematics, I claim to be the analog of Rutherford's position in science, the formal analog of "hardness" and quantitation.

The formalist position is, first of all, an expression of a belief that all mathematical truth can be reduced to, or expressed in terms of, word processing or symbol manipulation. Hence the close association of formalization with the idea of "machines" (Turing machines) and with the idea of algorithms. These embody purely automatic procedures, which require no thought, no perception, indeed, no external agency at all.

Second, the formalist position, that the universe of discourse needs to consist of nothing more than meaningless symbols pushed around by definite rules of manipulation, is exactly parallel to the mechanical picture of the phenomenal world as consisting of nothing more than configurations of structureless particles, pushed around by impressed forces.

The formalist position seems, on the fact of it, very attractive. For, by asserting that all truth is syntactic truth, it tells us that (1) we lost no shred of mathematical truth in the process of formalization, and (2) we are automatically guaranteed that mathematics is consistent. We pay for these benefits by giving up the idea that mathematics is "about" anything, i.e., that its propositions express percepts or qualities, but on the other hand we are *informally* free to interpret these propositions in any way we want. These are, of course, exactly the same attractions that the "hard" or quantitative sciences offer in the phenomenal world.

GÖDEL'S THEOREM

The celebrated Incompleteness Theorem of Gödel[3] effectively demolished the formalist program. Basically, he showed that, no matter how one tries to formalize a particular part of mathematics (Number Theory, perhaps the inmost heart of mathematics itself), syntactic truth in the formalization does not coincide with (is narrower than) the set of truths about numbers.

There are many ways to look to Gödel's Theorem. Indeed, the Theorem

2. S. C. Kleene, *Metamathematics* (New York: van Nostrand, 1952).
3. K. Gödel, *Monatshefte Math. Physik* (1931), 38:173–198.

itself has provoked an enormous literature, as might be expected. For our purposes, we may regard it as follows: *one cannot forget that Number Theory is about numbers*. The fact that Number Theory is about numbers is essential, because there are percepts or qualities (theorems) pertaining to numbers that cannot be expressed in terms of a given, preassigned set of purely syntactic entailments. Stated contrapositively: no finite set of numerical qualities, taken as *syntactical* basis for Number Theory, exhausts the set of all numerical qualities. There is always a purely semantic residue, that cannot be accommodated by that syntactical scheme.

Gödel's Theorem thus shows that formalizations are part of mathematics, but not all of mathematics. Mathematics, like language itself, cannot be freed of all referents and remain mathematics. Any attempt to do this (i.e., any attempt to capture *every* percept through a formalization of any finite set of percepts) must already fail in the Theory of Numbers.

On the other hand, Number Theory is still mathematics, still a system of inferential entailment in itself. It is only that it is not a purely syntactic system, not entirely a matter of word processing or symbol manipulation, independent of any external referent. In other words, Number Theory is not a closable, finite system of inferential entailment. These facts, as embodied in Gödel's Theorem, do not make us give up Number Theory as a part of mathematics nor even give up formalization as a strategy for studying certain kinds of mathematical systems. They express rather the limitations of formalization; it is not, as Hilbert thought, a *universal* strategy. If mathematics is a war against inconsistency, then that war is simply not as easily won as Hilbert believed.

COMPLEX SYSTEMS

The relation between Number Theory and any formalization of it concretely embodies certain features that bear essentially on the dualism with which we started, between "hard" or quantitative science (which asserts roughly that physics must be everything) and "soft" or qualitative sciences (which assert that physics is nothing).

The first thing to bear in mind is that both Number Theory and any formalization of it are both systems of entailment. It is the *relation* between them, or more specifically, the extent to which these schemes of entailment can be brought into congruence, that is of primary interest. The establishment of such congruences, through the positing of referents in one of them for elements of the other, is the essence of the *modeling relation*, which I have discussed at great length elsewhere.[1]

In a precise sense, Gödel's Theorem asserts that a formalization, in

Zone of Incongruencies
Define a Triangle as $\angle A + \angle B + \angle C = 180°$
or as △

which all entailment is syntactic entailment, is too *impoverished in entailment* to be congruent to Number Theory, no matter how we try to establish such a congruence. There are thus qualities pertaining to numbers, and to Number Theory, that are missed by any such attempt; hence any entailments in Number Theory pertaining to these unencoded qualities are likewise inaccessible in the formalization. It would thus require, at best, an infinite number of distinct formalizations to capture all the qualities, and hence, all the entailments of Number Theory, in terms of syntax alone.

This kind of situation is what I have elsewhere termed *complexity*.[4] In this light Gödel's Theorem says that Number Theory is more *complex* than any of its formalizations, or equivalently, that formalizations, governed by syntactic inference alone, are *simpler* than Number Theory. To reach Number Theory from its formalizations, or more generally, to reach a complex system from simpler ones, requires at least some kind of limiting process.

Simple
complex

Rutherford's position, as articulated above, can be rephrased as asserting that *every material system is a simple system*. Indeed, I have shown elsewhere that this position is just another form of Church's Thesis,[5] a direct assertion of the simulability (i.e., the purely syntactic character) of mathematical models of reality (i.e., of systems of *causal* entailments).

To a mathematical Rutherford, then, Number Theory would look *soft* relative to its formalizations, precisely because there are more qualities, and hence more entailments, in Number Theory than could be accommodated in terms of "hard" (i.e., syntactic) entailments. This is exactly why biology looks soft to a physicist, for example. I believe I need not belabor this situation, and the mistaken presumptions it manifests, any further.

On the other hand, let me now observe that the relation between Number Theory and a formalization of it can be iterated. That is: in the discussion so far, I have treated Number Theory as a "system," and formalizations as "models" of it. As we have seen, formalizations, being purely syntactic, have too little entailment to capture all the qualities of Number Theory itself. But as I have said, Number Theory is in itself a system of entailments, not only in itself a perfectly good mathematical system, but in many ways the very center of mathematics. Let us suppose we treat *it* as a model (i.e., as we treated formalizations before) and ask what *it* can model.

The question immediately arises: are there other mathematical formal-

4. R. Rosen, *Int. J. Gen. Systems* (1977), 3:227–232.

5. R. Rosen, *Bull. Math. Biophysics* (1962), 24:375–393; R. Rosen, in *The Universal Turing Machine: A Half-Century Survey*, R. Herken, ed., (Kammerer & Unverzagt, 1988), pp. 523–537.

isms too rich in entailment to be captured by Number Theory, and hence *more complex* than it? The answer, of course, is *yes;* in fact we can iterate this process, obtaining more and more complex formalisms, indefinitely.

At each stage of this iteration, a formalism appearing at that stage would appear "soft" with respect to any formalism appearing at an earlier stage. But of course, at no point do we exempt ourselves from entailment itself; quite the contrary.

On the other hand, a mathematical Hutchins might argue as follows: *because* Number Theory is "soft" relative to its formalizations (i.e., is more complex than they are), it is therefore immune to *mathematics* and must be studied by other means. This is tantamount to abdicating to syntax alone the right to call itself mathematics and declaring thereby that nonsyntactical modes of entailment fall outside the scope of mathematics. The material counterpart of this line of reasoning is to exempt complex systems from the domain of science precisely because they are complex. This view, it seems to me, is equally mistaken.

SUMMARY

As will be clear from the preceding discussion, it seems to me that the duality between "hard" or quantitative science and "soft" or qualitative science rests on an entirely false presumption. It is not in fact a question of Rutherford versus Hutchins, i.e., a question of doing physics or not doing science at all. It is rather a relative question, of simplicity versus complexity.

There is, as yet, no comprehensive investigation of the ideas I have sketched in the course of the discussion above; they are too new. But it seems that such ideas, or ideas like them, are necessary in many ways. I would in particular draw attention to the way such ideas ultimately rest on entailment alone, on systems of entailment in the material world (causal entailment) and in the world of formalisms or mathematics (inferential entailment), and on comparisons or congruences between such entailment systems. I have come to believe that the concept of entailment provides a reliable anchorage for the scientific enterprise itself, and I accordingly recommend it to your attention.

Chapter I

Prolegomena

THIS BOOK represents a continuation, an elaboration, and perhaps a culmination of the circle of ideas I have expounded in two previous monographs: *Fundamentals of Measurement and the Representation of Natural Systems* (henceforth abbreviated as *FM*) and *Anticipatory Systems* (abbreviated as *AS*). Both of these, and indeed almost all the rest of my published scientific work, have been driven by a need to understand what it is about organisms that confers upon them their magical characterisics, what it is that sets life apart from all other material phenomena in the universe. That is indeed the question of questions: What is life? What is it that enables living things, apparently so moist, fragile, and evanescent, to persist while towering mountains dissolve into dust, and the very continents and oceans dance into oblivion and back? To frame this question requires an almost infinite audacity; to strive to answer it compels an equal humility.

I A. What Is Life?

Ironically, the idea that life requires an explanation is a relatively new one. To the ancients, life simply *was;* it was a given; a first principle, in terms of which other things were to be explained. Life vanished as an explanatory principle with the rise of mechanics, when Newton showed that the mysteries of the stars and planets yielded to a few simple rules in which life played no part, when Laplace could proudly say "Je n'ai pas besoin de cet hypothèse"; when the successive mysteries of nature seemed to yield to understanding based on inanimate nature alone; only then was it clear that life itself was something that had to be explained.

From whence shall such explanation come? To which oracle shall I put the question? The first thought is: to that same mechanics, that same physics, which first exorcised life from the heavens and which has since plumbed the depths of matter, space, and time. Living things are surely material; they are manifestations of matter; surely then the secrets of matter must contain the secrets of life. Surely the physicist, who is concerned with matter in all of its manifestations, will have eagerly striven to translate insights about matter in general into corresponding insight about matter's greatest mystery.

Oddly enough, the physicist, qua physicist, has shown no such eagerness. The historical fact is that the phenomena of biology have played essentially no role in the development of physical thought or in the application of that thought. Why? Mainly, I think, because theoretical physics has long beguiled itself with a quest for what is universal and general. As far as theoretical physics is concerned, biological organisms are very special, indeed, *inordinately* special systems. The physicist perceives that most things in the universe are not organisms, not alive in any conventional sense. Therefore, the physicist reasons, organisms are *negligible;* they are to be ignored in the quest for universality. For surely, biology can add nothing fundamental, nothing new to physics; rather, organisms are to be understood entirely as specializations of the physical universals, once these have been adequately developed, and once the innumerable constraints and boundary conditions that make organisms special have been elucidated. These last, the physicist says, are not my task. So it happens that the wonderful edifice of physical science, so articulate elsewhere, stands today utterly mute on the fundamental question: What is life?

One of the few physicists to recognize that the profound silence of contemporary physics on matters biological was something *peculiar* was Walter Elsasser. To him, this silence was itself a physical fact and one that required a physical explanation. He found one by carrying to the limit the tacit physical supposition that, because organisms seem *numerically* rare in the physical universe, they must therefore be too special to be of interest as material systems. His argument was, roughly, that anything rare disappears completely when one takes averages; since physicists are always taking averages in their quest for what is generally true, organisms sink completely from physical sight. His conclusion was that, in a material sense, organisms are governed by their own laws ("biotonic laws"), which do not contradict physical unversals but are simply not derivable from them.

Ironically, ideas like Elsasser's have not had much currency with either physicist or biologist, although one might have thought they would please both. Indeed, in the case of the former, Elsasser was only carrying one step further the physicists' tacit supposition that "rare" implies "nonuniversal."

The possibility is, however, wide open that this supposition itself is mistaken. On the face of it, there is no reason at all why "rare" should imply anything at all; it needs to be nothing more than an expression of how we are sampling things, connoting nothing at all about the things themselves. Even in a humble and familiar area like arithmetic, we find inbuilt biases. We have, for instance, a predilection for rational numbers, a predilection that gives them a weight out of all proportion to their actual abundance. Yet in every mathematical sense, it is the rational numbers that are rare and very special indeed. Why should it not be so with physics and biology? Why could it not be that the "universals" of physics are only so on a small and special (if inordinately prominent) class of material systems, a class to which organisms are too *general* to belong? What if physics is the particular, and biology the general, instead of the other way around?

If this is so, then nothing in contemporary science will remain the same. For then the muteness of physics arises from its fundamental *inapplicability* to biology and betokens the most profound changes in physics itself. This situation is, of course, nothing new in physics; it happened when physics was mostly mechanics and had no apparent room for accommodating phenomena of electricity and magnetism; it happened again when the combined arsenal of nineteenth-century physics spent itself helplessly in assaulting phenomena of spectra and chemical bonding. But just as today's armies are equipped only to win yesterday's wars, we cannot expect contemporary physics to successfully cope with problems other than those with which it has already coped.

As we proceed, we will find a great deal of evidence, of many kinds, that leads to just such an unfortunate conclusion. And thus, we find another, basic reason why biology is hard; it is hard because we are fundamentally ill equipped. This is a far cry from merely being ignorant; it is rather that we are misinformed.

In any event, the task that physics has shirked devolves next onto biologists, perhaps properly so, since it is central to their every enterprise. What light, then, do biologists shed on the taproot of their own endeavors? In fact, precious little. Indeed, a rather strange and dreary consensus has emerged in biology over the past three or four decades. On the one hand, biologists have convinced themselves that the processes of life do not violate any known physical principles; thus they call themselves "mechanists" rather than "vitalists." Further, biologists believe that life is somehow the *inevitable* necessary consequence of underlying physical (inanimate) processes; this is one of the wellsprings of reductionism. But on the other hand, modern biologists are also, most fervently, evolutionists; they believe wholeheartedly that everything about organisms is shaped by essentially *historical, accidental* factors, which are inherently unpredictable and to which no universal principles can apply. That is, they believe that everything important about life is *not* necessary but contingent. The unperceived ironies and contradictions in these beliefs are encapsulated in the recent boast by a *molecular* biologist: "Molecular biologists do not believe in equations." What is relinquished so glibly here is nothing less than any shred of logical necessity in biology, and with it, any capacity to actually understand. In place of understanding, we are allowed only standing—and watching. Thus, if the physicist stands mute, the biologist actually negates, while pretending not to.

Thus, to ask the question "What is life?" is to find oneself standing essentially alone. But perhaps not entirely; there is yet another oracle to be consulted. That oracle is System Theory, which as yet speaks only in whispers. Insofar as it can be characterized, System Theory is the study of organization per se, divorced from material embodiment, as the form of a statue can be divorced from the marble or as cardinality can be divorced from the things being enumerated. This oracle will at least entertain the question but only when it has been transmuted to a new form: not what is special about life in terms of matter but what is special about it in terms of organization. This is good, because obviously organisms are, in purely material terms, of the greatest diversity. But it is not, by itself, good enough. Moreover, this oracle speaks not of laws or principles, as physics does; as yet it can speak only in parables.

Thus, the question remains, and it is still *the* question: What is life? It commands us to grapple with it and even allows us the luxury of choosing our own weapons for the struggle. But our armory is inadequate; wherever we look, some essential element is missing. Life is material, but the laws framed to describe the properties of matter give no purchase on life. Something is missing here, perhaps something essential for the understanding of matter in general, however much the physicists insist not. Biology has so far spent itself in cataloguing the endlessly interesting epiphenomena of life, but at the heart of it there is still only a gaping void. And the parables of the system theorist cannot as yet be incarnated in material reality. As I said, something is missing, something big, but it is hard to see even the biggest things when they are not there. We can only sense the void of its absence and try to fabricate what is necessary to fill it.

That is what the present book is about. It is about the creation and the application of an armory for a renewed assault on the question of questions: What is life? In the process, we shall find ourselves partly in the world of physics, constructing a language appropriate for a physics of "organized matter," a physics of *complex systems*. We shall also find ourselves partly in the world of the system theorist, developing a language appropriate to "material organization" and thereby clothing that world in a substance and coherence it has largely lacked. We will be in the world of the mathematician, in the world of formalisms and formalizations. And finally, of course, we will find ourselves in the world of biology, to see how far our armamentarium will take us in our struggle with the question: What is life?

I B. Why the Problem Is Hard

As a first step in our assault on the problem What is life? it will be well to get some idea of what we are up against. Specifically, we will try to understand what it is about the problem that has rendered it so refractory to the combined resources of our contemporary scientific wisdom. This will provide one way of sensing the shape of the void we need to fill and at the same time will help set the stage for our further, more technical developments, though we will not be able to reach a true answer to this question until we come to the end.

Let us begin by noting the very form of this question; we are asking *why*. We shall find ourselves asking "why" very often as we proceed. The answer to such a question (and indeed there are in general many ways to answer such a question) is to assert a "because." As we shall see abundantly later, to ask why is to enter the realm of causality, and to propose an answer is to posit something, to make a hypothesis. Although every physicist must believe in causality, this attitude toward positing a "because" was set long ago by Newton, whose proudest assertion was *hypothesis non fingo*. Indeed, as we shall see, causality in contemporary physics has evolved into a very different kind of thing than that originally envisaged by Aristotle, a thing geared essentially to deal with the question "what?" and to provide answers of the form "this."

One of the main reasons the fundamental question "What is life?" is so hard will turn out to be closely associated with such ideas. It will turn out that this question is really a "why" question in disguise, that we are really asking, in physical terms, why a specific material system is an organism, and not something else. Such questions are not congenial to contemporary science. To take a simple example: If we give a physicist, say, a clock, his or her interest will reflexively concentrate entirely on how it works, *never on how it came to be a clock*. We shall see below that the whole formal structure of theoretical science is geared toward the former question and away from the latter. Indeed, it will turn out that a large part of the work we need to do arises from exactly such elementary considerations.

But there are other possible reasons why the problem is hard. One of the most facile, frequently adduced, and fundamentally misleading reasons is to assert that *we simply do not know enough yet* to meaningfully approach the problem; we need more data. That is, difficulty merely presupposes ignorance, that we lack only an appropriate factual basis on which to proceed. It then follows precisely that, because the problem is hard, any attempt to deal with it is *premature*.

It is well to spend a moment discussing this possibility, since it is plausible on at least two counts. First, the fact is that biology itself is extremely young as a science. Indeed, the very word "biology" dates only from about 1800 or so. It is also true that biology has seldom been able to develop, from within itself, the instruments required for probing what actually goes on inside an organism. For example, however important it was to be able to scrutinize the intimate details of organic anatomy, biology had to wait upon the development of a completely independent physical science (optics) in order to fabricate and deploy so basic an instrument as the microscope. Likewise, the field of biochemistry, not to mention its offshoots of molecular biology and molecular genetics, could not have developed beyond the crudest of generalities without techniques for tracing, manipulating, and characterizing minute quantities of matter—techniques that became available only through developments in atomic and nuclear physics in the 1930s.

Thus, biology has generally had to parasitize other sciences in order to develop its own experimental techniques, and consequently, has barely begun to accumulate its fundamental data. Extrapolating from this historical picture, we may expect that, as new experimental techniques are generated elsewhere, entirely new biological probes, and correspondingly new data, presently unimaginable, await us in the future.

Therefore, just as it was premature to speculate on organic microanatomy before microscopes were available, or to speculate in 1920 on how proteins were synthesized, so too it may be premature to venture upon the much deeper question of "What is life?" today. And the argument can be strengthened by looking at the history of physics itself. Of what value were speculations regarding the source of atomic spectra, or chemical bonding, or the origin of stars in 1850? These were hard problems then, precisely because the factual basis to deal with them was lacking.

Such arguments cannot be discounted a priori, of course, but they can be considerably weakened by precisely the same historical considerations that give them their apparent weight. For one thing, the two most important ideas in contemporary biology, namely, Mendelian genetics and Darwinian evolution, were both premature, according to these views. For another thing, a fact or datum, by itself, is essentially meaningless; it is only the *interpretation* assigned to it that has significance. Thus, for example, one can literally see the rotation of the earth on any starry night; it has always been patently visible, but for millennia human beings did not know how to understand or interpret what they were seeing. Examples of such misinterpretation, which have retarded the development of science by centuries, can be multiplied without end in the history of science; in all these cases, it was the absence, not of data, but of imagination that created difficulty.

Finally, and most important, a fact or datum cannot, by itself, answer a question "why?"

To sum up: It may perhaps be true that the question "What is life?" is hard because we do not yet know enough. But it is at least equally possible that we simply do not properly understand what we already know.

Let us turn to another circle of historical ideas that bear on why biology is difficult. If we look back at the history of physics, with a view to understanding why it could develop as early and as rapidly as it did, we find two relevant sets of circumstances. First, there was the patiently accumulated data of astronomy, collected and tabulated for millennia. Second, an apparently unrelated circumstance: one could experiment directly with simply mechanical contrivances, with inclined planes, pulleys, and springs, and express the resulting data in the form of simple mathematical rules or laws. It was tacitly assumed that these *same* rules, revealed by simply laboratory experiments with simple bulk systems, held good throughout all of nature, at every level, from the very greatest to the very smallest. The story of Newton and the falling apple is perhaps apocryphal, but there is no doubt that his ideas about gravitation were based on an assumption of the universal validity of mechanical extrapolations whether the objects involved were planets, or terrestrial projectiles, or ultimate atomic particles.

This tacit belief in the unlimited uniformity of mechanical behavior, and the corresponding universality of mechanical laws, provided the absolutely essential nutrient that permitted theoretical physics to develop as it did. For it implied that we could study, on a convenient terrestrial scale, the same forces that moved the planets on the one hand and the atoms of the other. It implied that we could, through the simplest laboratory situations, simultaneously study *all* of inanimate nature, that our humble laboratories were proxies for the entire universe.

We now know, after three centuries, that this assumption of uniformity was entirely, hopelessly false. We cannot extrapolate from bulk matter to what goes on in an atom, or from either of these to events on galactic scales. But luckily, it was correct enough, for long enough, to establish a firm foundation of physical thought, a heritage of mathematical language, that could then be infinitely modified in detail and in interpretation, so that when the falsity of the basic premise was exposed, we did not need to jettison everything and start again. In retrospect, we can see that physics has, in this respect, been blessed with unbelievable good fortune.

In biology, however, the situation has been entirely different. From the physical point of view, even the simplest system one would want to call an organism is already inconceivably complicated. There are no biological counterparts of the inclined plane or pulley, the simple system that already manifests in itself the general laws we want to study. We cannot thus study organisms by inorganic proxy, at least not experimentally (see *AS*). The best we can do is to *dismember* an organism, to break it apart, and treat its parts in isolation as proxies for parts *in situ*.

These facts have decreed a very different historical development for biology vis-à-vis physics. They have also served to isolate biology for centuries from the rest of natural science, an isolation to which I have already alluded. It is in fact only in recent years, the years of "molecular biology," that any direct contact at all has become visible between the triumphant ideas of seventeenth-century particle mechanics and the properties of organisms. It is most ironic that today's perceived conjunction between physics and biology, so fervidly embraced by biology in the name of unification, so deeply entrenched in a philosophy of naive reductionism, should have come long past the time when the physical hypotheses on which it rests have been abandoned by the physicist.

The circumstances sketched above provide another complex of reasons why biology is hard and why the ultimate question of "What is life?" is perhaps the hardest of all. But there is one more circumstance that is not without an irony of its own. So far in this section, I have dwelt upon fundamental differences between organic systems and other kinds of material systems, differences that have, as we have seen, essentially set biology apart from the growth of other sciences during the past centuries, and hence, from the canons of explanation and understanding that seemed to work so well everywhere else. From these considerations, one would expect that such vast differences could themselves be quantified, that there should at least be some explicit, tangible, categorical test for distinguishing between a material system that is an organism and one that is not.

Indeed, many people have tried very hard to produce such criteria for separating the quick from the dead. Put briefly, they have all failed. This fact is most significant in itself, but perhaps of even greater significance is the further fact that we can somehow *know*, with certainty, that they have indeed failed.

These facts (let us assume that they are such for a moment) mean that we do not even know what biology is *about,* in the same sense that we know what mechanics is about, or what optics is about, or what thermodynamics is about. We thus do not know the scope or the domain of biology, for it has as yet no objectively definable bounds. In place of these, we have only a tacit *consensus,* which we somehow all seem to share, much as the native speakers of a language share the ability to discern its idioms. It is precisely against this tacit consensus that we judge the efficacy of a purported definition of organism, and it is by virtue of this consensus that we know all such definitions have in fact failed. They have failed because all of them, in one way or another, present to us as organisms objects that our consensus rejects, and equally bad, it classifies as nonorganisms systems that our consensus accepts.

The crux of the matter is that, when one tries to embody our recognition criteria for organisms in an explicit list, we find nothing on that list that cannot be *mimicked* by, or embodied in, some patently inorganic system. That is, there is no property of an organism that cannot at the same time be manifested by inanimate systems. Conversely, a dead organism is as inanimate as anything.

Indeed, a whole approach to biology, which used to be called *biomimesis,* was based on this observation. This approach might better be called "artificial biology." The idea was that a material system that could be made to embody "enough" properties of organisms would then *automatically be* an organism. Thus, people studied the motility of oil drops in salt solutions, the irritability and excitability of artificial membranes, the spontaneous division of growing droplets, and a host of other such phenomena. These ideas persist today in at least two areas: experimental approaches to "origin-of-life" and, in an entirely different material context, in robotics. Exactly the same idea is embodied in "Turing's Test"; a "machine" that has enough specific attributes of intelligence must *be* intelligent. From this an entire industry has arisen, devoted to the study of "artificial intelligence."

However we arrive at our consensus concerning what constitutes an organism, we do not implement it by checking off against a list. It may help to regard it rather as a form of "pattern recognition," but this only raises the question of what the pattern is that is being recognized. In any event, the fact remains that biology is not defined by, or characterized by, any discrete list of attributes that collectively reflect the consensus lying at the root of it. There is thus, in particular, no way we can impart that consensus to a "machine" in the form of a program. And as we shall see (see chapters 8–10), this fact itself, when appropriately articulated and judiciously combined with some of the other matters raised above, will take us a surprisingly long way.

I have already mentioned the word "machine" twice in the course of the past few paragraphs. It is a word that comes up often in contemporary biology, especially since almost all biologists account themselves as *mechanists*. Indeed, the protean concept of the machine, which in its way is as ill defined as that of organism itself, has dominated the inner world of biologists for far too long. The specter of the machine constitutes yet another reason why biology is hard. Let us then proceed to discuss it a bit, at least in a preliminary way; I shall have much more to say about it as we proceed.

I C. The Machine Metaphor in Biology

As we have just seen, one of the reasons biology is hard is that no one can say what an organism is. It is, however, all too easy to say what an organism is *like*. In itself, this is not a bad thing to do; trouble arises when one substitutes the latter for the former.

The earliest and most mischievous instance of this kind of substitution goes back to René Descartes. Apparently, Descartes in his youth had encountered some realistic hydraulic automata, and these had made a great impression on him; he never forgot them. Much later, under the exigencies of the philosophical system he was developing, he proceeded to turn the relation between these automata, and the organisms they were simulating, upside down. What he had observed was simply that automata, under appropriate conditions, can sometimes appear lifelike. What he concluded was, rather, that *life itself was automaton-like*. Thus was born the machine metaphor, perhaps the major conceptual force in biology, even today.

Descartes took this fateful step with only the haziest notion of what a mechanism or automaton was (Newton was still a generation away), and an even dimmer notion of what an organism was. But Descartes was nothing if not audacious. Descartes' conception was in fact perfectly timed; the triumphant footsteps of Newtonian mechanism were right behind it; the apparently unlimited capabilities of machines were already on their way toward a complete transformation of human society and human life. Why indeed should the organism not be a machine? There is no denying the many powerful allures encapsulated in the Cartesian metaphor; it hath indeed a pleasing shape.

Aside from its purely scientific and methodological implications, the psychological appeal alone of the machine metaphor to biologists over the years has been immense. We have already noted the profound isolation of biology from the dramatic developments in physical science since the time of Newton. The idea of the organism as machine permitted at least a vicarious contact with all this; it was plausible, easy to grasp, and above all, *scientific;* it showed a way around hazy metaphysical notions like intentionality, *telos,* finalism, and the like, which have always plagued biology, but which physics had apparently discarded for centuries past.

Today, I dare say that the molecular biologist, above all, regards this field as representing the ultimate incarnation of the Cartesian automaton. The molecular machine has now displaced all earler metaphorical images of the organism as clockwork, as engine, as chemical factory, as servomechanism, and as computer. Genetic engineers, who are the molecular biologists turned technologues, habitually regards their favorite organism, *E. coli,* as a simple vending machine; insert the right token, press the right button, and the desired product is automatically delivered, neatly packaged and ready for harvest.

But above all, the machine metaphor (supported, of course, by the corpus of modern physics) is what ultimately drives, and justifies, the reductionism so characteristic of modern biology. For whatever else a machine may be, it is a composite entity; it is made up of parts. The way these parts interact to produce the machine's characteristic behaviors constitutes its *physiology;* the way the machine is assembled from these (very same!) parts accounts for its *origin.* The way to learn about these things is merely a matter of dissection or fractionation, to dismember the machine into its constituent parts and characterize them individually as independent subsystems. That is what we do with a machine, be it a watch, or a piece of electronic equipment, or an automobile. It is also what a physicist does with a stone, a molecule, or an atom. All of these procedures embody the first rule of scientific analysis: separate mixtures into pure substances, devolve the properties of the whole onto the properties of the parts.

For the machines we build for ourselves, there is indeed a "set of parts" into which the machine can be resolved; these are obtained essentially by reversing the process by which the machine was fabricated in the first place. These parts can be separated and characterized, without losing any information pertinent either to the physiology or the reconstruction of the machine itself. The belief in reductionism, buttressed precisely by the machine metaphor, extrapolates these facts back to the entire universe; there is *always* a set of parts, into which *any* material system (and in particular, any organism) can be resolved, *without loss of information.* Specifically, the properties of these parts, considered in isolation, collectively entail the properties of the fully articulated system; moreover, they even imply how the parts themselves are to be articulated. In short, the posited set of parts constitutes the natural analytic subunits for simultaneously resolving two *entirely different problems:* how the system is to be constructed (the fabrication problem) and how the system actually works (the physiological problem).

Of course, not every way of decomposing or fractionating a machine will give us a set of parts with these happy properties. Taking a hammer to a watch, for example, will give us a spectrum of parts all right; these may be separated and characterized to our heart's content, but only by a miracle will they tell us either how a watch works or how to make one. This is because two things have happened: application of the hammer has *lost* information about the original articulated watch, and at the same time, it has *added* irrelevant information about the hammer. What the hammer has given us, then, is not so much a set of parts as a set of artifacts.

Nevertheless, the machine metaphor is what whispers to biologists that a set of parts exists; they whisper to themselves that they have found them, with respect to physiology if not fabrication. And yet, we must realize that physicists have long known about very simple material systems, whose "physiologies" cannot be analyzed in this way, even in principle. One cannot, for instance, solve a three-body problem by breaking a three-body system (e.g., the earth, sun, and moon) into three one-body systems or even into a two-body and one-body system. Already in this case, we must draw one of two conclusions: (1) there *is* no set of parts, or analytical subunits, from whose properties in isolation the physiology of the intact system can be reconstructed, or (2) if there are such parts, they are not the obvious ones; they must rather be of a completely different and novel character. Moreover, in this last case, we cannot expect such parts to solve both the physiological and the fabrication problem simultaneously. Instead, entirely separate analyses are required, as befits the two entirely different kinds of problems to be solved.

To sum up: the role of the machine metaphor in biology today is as follows. First, it assures biologists that their subject is an analytical one, because it asserts that any machine is a set of parts. Second, it assures them that the *same* set of parts will solve all problems of fabrication and of physiology simultaneously. Third, it assures them that nothing happens in biology that is outside the ken of the physical universals (or rather of those fragments of physical universality necessary for the understanding of machines). As to the parts themselves, biologists used to think that they were cells, but today they are molecules. And if biology is hard, it is simply because there are so many parts to be separated and characterized.

This last paragraph encapsulates, I think, the working biologist's view of reductionism. If the machine metaphor, which is its primary mainstay, is even a little bit wrong, then this metaphor itself makes biology infinitely harder than it needs to be. It makes biology objectively harder, because it transmutes biology into a struggle to reconcile organic phenomena with sets of constituent fragments of unknown relevance to them; it makes biology subjectively harder because biologists have committed themselves to the analysis rather than to the organism. The question "What is life?" is not often asked in biology, precisely because the machine metaphor already answers it: "Life is a machine." Indeed, to suggest otherwise is regarded as unscientific and viewed with the greatest hostility as an attempt to take biology back to metaphysics.

This is the legacy of the machine metaphor. I hope to convince the reader, in the course of the present work, that the machine metaphor is not just a little bit wrong; it is entirely wrong and must be discarded.

Chapter 2

Strategic Considerations: The Special and the General

I HAVE ASSERTED, several times in the course of the previous discussion, that "something is missing" from the resources we bring to bear on the question "What is life?" The present section is concerned with making this assertion sharper. In the course of doing this, I shall describe a few of the numberless examples of how, in the past, science and mathematics have stumbled into, or more often, been pushed by the pressure of circumstance, into reluctant realization that "something was missing" from their inherited worldview. Such occasions have almost invariably been associated with pain, and the remedies dramatic. Indeed, as true in science as in any other field of human endeavor are the words that George Bernard Shaw puts in the mouth of Andrew Undershaft: "You have learned something. That always feels, at first, as if you had lost something."

2A. Basic Concepts

I shall begin with a brief consideration of the concepts of generality and universality. Specifically, what does it mean to say that a theory is general or universal? Or that a proposition is generally or universally true? It is true that, say, Set Theory is more general than topology, or that Lattice Theory is more general then Group Theory, or that mechanics is more general (universal) than thermodynamics? Or that physics is more general (or less general) than biology? What sense, if any, can we give to such questions?

Let us look at the first question: is set theory more general than topology? At first sight, this question admits contradictory answers. For a topological space is a set, plus additional structure. We might say that the additional structure we need to specify makes topology more special. On the other hand, every set *is* a topological space, if we imagine it equipped with the trivial (i.e., very special) topology. From this point of view, we could equally say that the topological space is the more general concept (since it generalizes the *topological* structure that inheres in every set).

In fact, the question I have raised does not admit resolution at this level; the decision point lies somewhere else. It lies in the fact that, once we have imposed an additional structure on a set, be it of an algebraic, topological, or any other character), the *mappings* we use to compare one such structure with another must, of course, respect that structure. In our case, topological structures are compared by using mappings that are *continuous*. The set of *continuous* mappings from one topological space to another is thus generally much smaller than the totality of unrestricted mappings between their underlying sets. It is at this level that the restrictive nature of the additional structure manifests itself. And on this basis, we have grounds for asserting that set theory is the more general.

On similar grounds, we can argue that mechanics is more general than thermodynamics. Or rather, we can argue that the *formalism* of mechanics is more general than is that of thermodynamics. On this basis, we can assert then that mechanics is the more *universal* theory; I shall have much more to say about this subsequently (see chapter 4).

Comparing *different* kinds of mathematical structures, as for example, a group structure with a lattice structure, requires still another set of ideas, if we wish to talk about the generality of, say, group theory versus lattice theory. What we need to do in this case is to produce one kind of structure in the context of the other, and conversely. For instance, the set of all subgroups of a given group forms a lattice; the set of all automorphisms of a lattice forms a group. If we could do this in such a way that the resulting pair of operations were inverses of one another (e.g., if we could assert, which we cannot, that "the group of lattice automorphisms of the lattice of all subgroups of a group is the group itself" and the corresponding inverse proposition), then we would have grounds for saying that the two theories are equally general; if we could say one and not another, we could rank one theory as more general; if we can say nothing at all, then the generality of the two formalisms cannot be compared; they are of *different* generality.

These rather vague ideas regarding generality serve as a kind of taxonomic index, according to which formalisms may be classified. As we shall see later (see section 3H below), when I come to talk about modeling of one kind of system in another, they actually do much more than this. But for the moment, it is sufficient to note that generality rests on the idea of *inclusion;* a formalism (e.g., set theory) is more general than another (e.g., topology) if there is some kind of proper inclusion that can be established somewhere (in this case, the set of continuous maps between toplogical spaces is a proper subset of the set-theoretic maps between their underlying sets).

In any case, passage from general to special corresponds to the imposition of additional structure, additional conditions not satisfied by the typical (or to use a more modern word, by the *generic*) member of the more general class. Conversely, passage from special to general involves the removal or waiving of such special conditions.

Let us now consider a few well-known historical examples.

2B. From General to Special

From the foregoing, we can see that given any formalism, say the Theory of Groups, we can *generalize* it by simply throwing away some of its defining structure. For instance, if we throw away the group axiom requiring every element to have an inverse, we get a more general structure called a *semigroup* or *monoid.* Here the inclusion on which the assertion of generality rests is patent; every group is a fortiori a semigroup, but not conversely. Indeed, it is not hard to see that, in some precise sense, it is most unusual *(nongeneric)* for a semigroup to *also be a group* or even for an *element* of a given arbitrary semigroup to possess an inverse.

Conversely, the way to pass from a more general situation, (e.g., semigroups) to a more special one (e.g., groups) is to *add conditions.* Namely, we must impose further structure, which serves to discriminate between what is special and what is not. Obviously, the more restrictions we impose, the smaller will be the class of systems that satisfy them all. On the other hand, because there are more conditions to be satisfied, we can usually say more things, and deeper things, about those mathematical objects that do satisfy them. In mathematics, we must strike a balance between these two conflicting aspects of specialization; namely, we must impose enough conditions to allow interesting theorems about the objects satisfying them, but not so many conditions that the class of such objects is too small to itself be interesting. This kind of balance is in fact the mathematical paradise, where one can prove interesting things about interesting objects. We have, of course, seen this before (see section 1A above); it is why physics regards biology as "too special" to be of physical interest.

We have seen that, *given* a formalism, we can specialize it by simply adding appropriate further conditions. But what if we do not exactly have a formalism to begin with? This kind of situation has occurred many times in mathematics, and in its most serious form has been responsible for most of the so-called Foundation Crises that mathematics has experienced over the years. In these cases, some kind of apparently allowable operation or procedure was "too general," in the sense that its unrestricted use implied terrible things like "$1 = 0$." The question became one of *how* to restrict or circumscribe the procedure, to specify conditions under which it became safe.

A classic example of this situation is the following. The ordinary arithmetic operations (addition and multiplication) are essentially *binary* operations; they apply to pairs of numbers and yield a definite sum and product respectively. They can immediately be extended from pairs of numbers to any *finite* set of numbers; we can give a unique and unambiguous meaning to expressions like

$$r_1 + r_2 + \ldots + r_n; \qquad r_1 \times r_2 \times \ldots \times r_n.$$

Every child knows this.

A whole new mathematical universe was glimpsed through removal of the restriction of finiteness, i.e., by *generalizing* from finite arithmetic to infinite sums and infinite products. In the hands of people like Euler, the Bernoullis, and others, a host of new and beautiful relations were generated; for instance

$$1 + \frac{1}{9} + \frac{1}{25} + \ldots + \frac{1}{(2n-1)^2} + \ldots = \frac{\pi^2}{8};$$

$$\sqrt{\frac{1}{2}} \cdot \sqrt{\frac{1}{2} + \frac{1}{2}\sqrt{\frac{1}{2}}} \cdot \sqrt{\frac{1}{2} + \frac{1}{2}\sqrt{\frac{1}{2} + \frac{1}{2}\sqrt{\frac{1}{2}}}} \ldots = \frac{2}{\pi};$$

$$\frac{2}{1} \cdot \left(\frac{4}{3}\right)^{1/2} \cdot \left(\frac{6}{5} \cdot \frac{8}{7}\right)^{1/4} \cdot \left(\frac{10}{9} \cdot \frac{12}{11} \cdot \frac{14}{13} \cdot \frac{16}{15}\right)^{1/8} \ldots = e.$$

It seemed as if the ordinary laws of finite arithmetic extended benevolently into this new, infinite realm: products distributed over sums; the result of an infinite summation or multiplication was independent of how it was parenthesized, etc.

The mathematical masters never (at least publicly) encountered any difficulties in manipulating infinite sums and products. But in the lesser hands that strove to emulate them, patently absurd results began to appear. The worst of them is this:

$$1 + (-1) + 1 + (-1) + \ldots$$

It allows us apparently to conclude that any integer is equal to any other.

Obviously, the world of unrestricted infinite sums and products is somehow *too* large, *too* general. *Sometimes* we can safely navigate in this world, but when? We could, of course, stay completely safe and never enter this world, but that would be too drastic and unacceptable; who would want to renounce beautiful results of the kind we exhibited above? The question became then how to *restrict* ourselves to those infinite sums and products that "make sense" and avoid those that do not.

The resolution of this crisis was given by Cauchy, who in 1805 introduced the necessary concept, *convergence*. He even gave an effective criterion for convergence, which anyone can apply. Thus, the original generalization from finite arithmetic to infinite had to be re-restricted in order to be meaningful; we need, not the universe of *all* infinite sums, but rather all those that satisfy the additional property that they *converge*. And we may note that, as usual, it is *nongeneric* for an infinite sum or product to also converge.

Historically, Cauchy's conception of convergence, which gives sense to limiting operations, and the associated concept of *continuity* (mappings that, in some precise sense, commute with these limiting operations) provided one of the basic cornerstones for what is today called *topology*. A topological space can be regarded as a minimal formalism in which convergence and continuity can be manifested. We may note, for future reference, that present-day topology has other historical antecedents, coming from geometry and dominated by quite different ideas related to congruence. In a way, it is the clash between these two distinct inheritances of topology that has given rise to the Theory of Categories; see section 5I below.

We are, at the moment, still in a Foundation Crisis, arising from the paradoxes inherent in Cantorian Set Theory. These paradoxes (or at least, the ones we know about; there is no guarantee that others may not be lurking), which were discovered around the turn of the century, clearly arise from an unrestricted (too general) use of concepts like membership and elementhood. Particularly suspect is the *reflexive* use of these concepts. As with the earlier problems associated with infinite sums and products, *sometimes* these concepts are meaningful and proper; sometimes they are not. Suffice it to say here that, as yet, no set-theoretic Cauchy has emerged to settle the situation, to find the level (or at least *a* level) of generality that is just right for Set Theory.

2C. From the Special to the General

In the preceding section, we saw some instances where mathematics has proceeded from the general to the special. Most of them were hygienic or therapeutic; in the absence of other exigencies, mathematicians always strive to go the other way, from what is special to what is general. Mathematics seeks the smallest set of conditions under which mathematical truth obtains, because this is the most elegant, the most parsimonious, the most illuminating, in short, the most beautiful.

Sometimes, however, the normally congenial act of generalization in mathematics is associated with great pain. The classic example is the discovery of the "non-Euclidean geometries," dating from about 1820. For over a millennium before that, "geometry" meant one thing; it meant what was between the covers of Euclid's *Elements*. The impact of Euclid upon mathematics in those days is hard to appreciate today; Euclid was the touchstone, the one part of mathematics that had attained absolute perfection. It embodied, in the truest way, the Platonic ideal of mathematics itself. And to use the phrase "non-Euclidean geometry" was at that time simply the most blatant contradiction in terms.

The crisis surrounding geometry that mushroomed in the ninteenth century grew from a single small blemish in the *Elements*, which the Greeks themselves, with their keen eye for such things, had already perceived. The blemish concerned the notorious Postulate of Parallels. The Greek post-Euclidean geometers did not doubt its "truth"; they merely felt it should be a theorem, because, with its potential for the unrestricted prolongation of lines, it was not as "self-evident" as a postulate should be. With their keen nose for aesthetics, they believed that the Parallel Postulate was in fact, redundant, that it was derivable from the other Euclidean axioms and postulates.

As it turned out, of course, the Parallel Postulate was not redundant; it was independent of the rest of Euclidean geometry. Thus, one could get different "geometries" simply by replacing Euclid's postulate with another. Hence, Euclidean geometry lost its unique character; it became only one geometry among many; it became *special*. In fact, it turned out that there was a single number (curvature) that characterized these geometries; Euclidean geometry was flat (curvature = 0); the Lobatchevskian or hyperbolic geometries were those with negative curvature; the Riemannian or elliptic geometries had positive curvature.

This kind of generalization (from zero curvature to nonzero) was of course highly traumatic under the circumstances and in fact was one of the wellsprings of the urgent concern with axiomatics (especially, of consistency of axiom systems), which fed into the Foundation Crises we have already mentioned. It was no longer, for example, even clear what "geometry" was any more. Within less than a half century, geometry had gone from being the most secure part of mathematics to the most insecure. It was a situation in which mathematicians were, for once, not seeking to generalize but were forced to do so.

At roughly the same time, very similar developments were occurring in theoretical physics. In mechanics, it had long been known that the equations of motion of a Newtonian system of particles (at least, a conservative system) were invariant under the Galilean group. This meant that different observers, moving relative to the system they were observing, and to each other, would see the same equations of motion, even though their measurements of position and momentum of particles in the system, tainted by their own motions relative to the system, would be different. In 1905, Einstein pointed out that these same Galilean transformations do not leave Maxwell's equations invariant. Thus, different observers of electrodynamic phenomena would not only come up with different data; they would also come up with different laws of motion. This, to Einstein, was contrary to experience. He showed that Maxwell's equations were invariant to a different group, the Lorentz group. The elements of this group contain a *parameter c,* and in fact, the Lorentz group collapses to the Galilean in the limit $c = \infty$ (or equivalently, when velocities v are small enough so that the ratio v/c is essentially zero). This parameter c was, of course, identified with a fixed (large) number, the velocity of light in vacuum. On plugging the Lorentz group back into mechanics, Einstein was able to draw a number of breathtaking conclusions, which are by now familiar to all.

This *Special Relativity* of Einstein constitutes a generalization of Newtonian mechanics, which came from a completely unexpected quarter. In replacing the Galilean group by the Lorentz group, we are actually doing much the same thing as changing geometries, replacing a flat or Euclidean world ($1/c = 0$) with a curved one; this kind of idea was later extended, by Einstein himself, in the passage from Special to General Relativity.

At roughly the same time (in 1900) another kind of generalization of Newtonian mechanics was in the works, stemming from Planck's discovery of the quantum of action. The nature of this generalization was not, however, fully appreciated for another quarter century, until the development of wave mechanics by Schrödinger, and the mathematically equivalent but very different-looking quantum mechanics of Heisenberg. In particular, the Uncertainty Relations on which quantum mechanics was based contained a (small) parameter h (Planck's constant), and quantum mechanics collapsed back to the Newtonian if we put $h = 0$ (i.e., all observables *commute*).

In fact, this generalization of Newtonian by quantum mechanics is closely related to one that had appeared more than three centuries earlier, namely, the generalization of geometric optics by the wave optics of Huyghens. By the middle of the nineteenth-century, W. R. Hamilton had established his Mechano-Optical Analogy (see *AS*), which allowed him to recast classical Newtonian mechanics in terms of a minimum principle (Least Action), and in the process established an exact dictionary between mechanics and geometric optics. By pursuing this analogy one step further and asking for the generalization of *mechanics,* which corresponded to the Huyghens' generalization of geometric optics, he would have discovered Schrödinger's equation; that is the way Schrödinger himself did it, a very long time later.

Let us look at one further instance of generalization, which we will consider in a mathematical context, but which historically has ties to engineering, and to quantum mechanics as well. These are associated initially with the name of Heaviside. At a time when mathematicians considered discontinuities to be pathological, and when they also thought they knew all about linear systems, Heaviside proposed using discontinuous signals (e.g., step functions) as an entirely novel probe of their behavior. Even worse was his introduction of derivatives of these discontinuities, to obtain things that mathematicians did not consider functions at all. Heaviside's revolutionary approach was thus completely discounted as ugly and as flying in the face of all accepted standards of rigor. His ideas received some prominence when they were adopted by the physicist Dirac, in the course of his development of an elegant formalism for the quantum theory itself (which is as linear as one could want, and where discontinuities are of the essence). However, even so great and so knowledgeable a mathematician as von Neumann completely dismissed Dirac's approach; his remarks are worth quoting directly:

> The method of Dirac . . . in no way satisfies the requirements of mathematical rigor, even if these are reduced in a natural and proper fashion to the extent common elsewhere in theoretical physics . . . the method . . . requires the introduction of "improper" functions with self-contradictory properties . . . the correct structure need not consist in a mathematical refinement and explanation of the Dirac method, but . . . requires a procedure differing from the very beginning. . . .

Nevertheless, within a little over a decade, these supposedly self-contradictory objects were shown to be perfectly rigorous and respectable. They were identified not with functions but with linear functionals, essentially with definite integral operators on linear spaces of functions. Ordinary ("proper") functions turn out to be a very special case, and the more general objects (originally called *distributions*) are identifiable with limits of sequences of these "proper" functions, just as real numbers are limits of sequences of rationals. Thus, from this point of view, distributions *generalize* ordinary functions; the generalization embeds ordinary functions in a new and larger universe, in which they are nongeneric indeed.

The repeated association of limiting processes with generalization of something, of which we have seen a number of instances so far, is itself a very general thing. Limiting processes are a gateway that can take us from a given world to a generally much larger world (since there are more sequences than elements), as we have already seen. The elements of this larger world may have new and different properties from those with which we started; properties generic in the large world but vacuous in the small one that gave rise to them. These remarks should be kept in mind, especially in light of what I have said about the relation of physics and biology. I will not use them for a long time (see chapter 10), but they should not be forgotten.

2D. Induction and Deduction: A Preliminary Note

The foregoing remarks regarding generalizations and specializations illustrate the kinds of ideas we shall need later, when we come to compare biology and physics, organisms and machines. They are, however, a far cry from what most people might expect a discussion of generals and particulars to be about, namely, about induction and deduction. Induction, roughly, seeks to establish general (i.e., quantified) propositions on the basis of *instances;* deduction, conversely seeks to establish instances in terms of quantified or general propositions. Such ideas turn out to have some relation, albeit remote, to the matters discussed above; however, since we shall need some ideas pertaining to induction later, it seems appropriate to say a few words about these issues here. Induction in particular is a very old problem, and the following remarks are intended to be in no way exhaustive.

To fix ideas, let X be some set, whose elements will be called *instances.* Let us further suppose, for simplicity, that we can enumerate the elements of X in some effective way; i.e., we can write

$$X = \{x_1, x_2, \ldots , x_n, \ldots \}.$$

We shall further denote by P some predicate or property that may or may not be manifested by a particular instance x_i in X; if the property P is possessed by the instance x_i, we shall write $P(x_i)$.

Let us denote by x an indeterminate or variable ranging over X, sometimes called a *free* variable. The expression "$P(x)$" is not itself an assertion but becomes one when instantiated by replacing the free variable x by a specific instance x_i in X.

The *universal quantifier* $\forall$ is a way of making a general assertion about the set or universe X. It is attached to the indeterminate x, and the result conjoined with $P(x)$. The resultant expression

$$(\forall x)P(x)$$

(in words, "for all x, or for any x, or for every x, x has the property P; $P(x)$ is true)" clearly tells us something about X itself. The expression can be *interpreted* as a string of conjunctions:

$$(\forall x)P(x) \; = \; P(x_1) \vee P(x_2) \vee \ldots \vee P(x_n) \vee \ldots \ldots$$

This is an example of a *general proposition* on X.

Clearly, we can always infer the *particular* $P(x_i)$ from the general proposition $(\forall x)P(x)$. This is *deduction;* in this form, one can see why deduction is regarded by many as trivial and why deductive sciences like mathematics tend to be despised by such people. However, such views completely miss the point; in mathematics, for example, the whole art is bringing a situation to a point where the final deducation *is* trivial. But that art itself is the antithesis of triviality; on the contrary, it is often of the most exquisite creativity and beauty.

The problem of *induction,* on the other hand, is a complementary attempt to go the other way; it is an attempt to establish a general proposition $(\forall x)P(x)$ on the basis of particular instances $\{x_{i_1}, x_{i_2}, \ldots, x_{i_r}$ in X for which it is known that $P(x_{i_k})$ holds.

Put another way: let S be a subset (a set of *samples*) of X, and let s be a free variable that runs only over the subset S. When we can conclude, from a knowledge that

$$(\forall s)P(s),$$

we must also have

$$(\forall x)P(x)?$$

That, roughly, is the general problem of induction; the effective passage from particulars (on S) to general on X).

Obviously, without further structure, the problem of induction cannot be solved in general; that much was originally pointed out by Aristotle, if not his predecessors, and elevated by Hume into a complete rejection of empiricism. By "empiricism" here, we mean the establishment of general truth by judicious sampling, so that what happens on an appropriately chosen sample set S can in fact be extrapolated to all of X.

The further structure necessary to deal with inductive problems effectively lies, of course, in the property P with which we are dealing. If P is simply any old property, then we cannot extrapolate from any sample set S to the whole set X, and the problem is indeed unsolvable by sampling. But if P, as a property, itself manifests what we may call *contagion,* so that the truth of $P(x_i)$ itself implies the truth of $P(x_j)$ for some other x_j's in X, then the problem of induction can be dealt with. Certainly in science, and in mathematics as well, we seldom deal with entirely arbitrary properties. Continuity, for example, is a contagious property, in the precise sense that what is true *at* a point remains true *near* the point.

The most vivid mathematical example of such contagion is embodied in what is correctly called *mathematical induction.* Roughly, induction in this sense requires only two things of a property P: (1) $P(1)$; and the crux of it, (2) for an arbitrary integer n, $P(n)$ *implies* $P(n + 1)$. The conclusion is: under these conditons, $P(n)$ is true for every integer.

Mathematical induction is, in fact, all we need to generate the whole of Number Theory from the existence of the number "1", and the ability to "add 1" to any integer. This indeed is the substance of the Peano Axioms for arithmetic, in which mathematical induction is the only inferential procedure available.

This idea will turn up again and again in our subsequent discussion; as we shall see, it permeates both science and mathematics in profound and insufficiently appreciated ways. I shall have much more to say about these matters at the appropriate time; see especially the discussion of *recursion* and its consequences, starting in section 4C below.

The efficacy of induction, the demonstration of general truths by extrapolating from samples, depends on the characteristics of the *property P* we are looking at. If, as is the case with mathematical induction, the property is contagious enough, then a sample of one is sufficient. Therein lies its strength.

2E. On the Generality of Physics

As already noted, it has been the prevailing sentiment in science today, as it has been for centuries past, that physics is the general, and hence, that biology is only a particular. In the present section, I will try to clarify what this assertion means.

In some ideal sense, of course, this assertion about physics is trivially true. Ideally, the *aspiration* of physics, its dream, is to encompass material nature in all of its manifestations. Organisms, as a part of material nature, clearly fall within this compass. Hence, from such an ideal perspective, biology indeed becomes part of the physical whole.

But I am not, in fact, addressing this ultimate ideal. When I use the word "physics," I am talking about *contemporary* physics; physics as it exists now, today, embodied concretely in all the books and journals reposing in the physics sections of all the libraries in the world. It is a very different matter to suppose, as reductionism requires, that biology is a particularization or specialization of *contemporary* physics.

To assess the "level of generality" of contemporary physics, or indeed of any other scientific or mathematical discipline, in any kind of absolute terms is an extremely difficult thing. It raises in fact a metatheoretic question; a question *about* the theory, not a question *within* the theory. Intuitively, the "level of generality" of a theory characterized the class of situations with which the theory can cope, the class of phenomena it can in principle accommodate. How, if at all, can such a thing be measured?

It is instructive, in this regard, to look at the Theory of Numbers in pure mathematics, where the situation is much more under control. Number Theory has historically been plagued with conjectures (really inductions, based on limited experience or sampling with small numbers), which no one has ever been able either to prove or produce a counterexample (disprove). Is Fermat's Last Theorem a theorem? How about the Goldbach Conjecture, that every even number is the sum of two odd primes? Is Number Theory general enough, even in principle, to cope with these very specific situations?

The situation is made even more interesting as a result of Gödel's celebrated work on undecidability in Number Theory, which we shall see much more of as we proceed. In brief, Gödel showed how to represent assertions *about* Number Theory *within* Number Theory. On this basis, he was able to show that Number Theory was not finitely axiomatizable. In other words: given any finite set of axioms for Number Theory, there are always propositions that are in some sense theorems but are unprovable from those axioms (unless, of course, the axioms are inconsistent to begin with—in which case everything is a theorem). The conclusion here is that *every* finitely axiomatized system of Number Theory is too special, in some abstract, absolute sense. But there is no way of telling whether a specific assertion or conjecture about numbers is provable, or disprovable, or undecidable (unprovable) within such a system.

If this is already the situation in Number Theory, how much more complicated to ask similar questions about physics. But that is exactly the question raised by reductionism; it is an assertion, or conjecture, or belief, pertaining to the generality of contemporary physics itself. And indeed, it is not a conjecture based on any *direct* evidence (as, say, Goldbach's Conjecture in Number Theory is), but rather on indirect (circumstantial) evidence, insofar as evidence is adduced at all. In short, it rests on *faith*.

As we have seen, generality is hard to assess in absolute terms. It is not as hard to assess *relative* generality, the generality of one theory with respect to another; I gave examples of this in earlier sections. In the present context, it is instructive to compare the level of generality of contemporary physics with that of physics as it was, say 100 years ago.

Indeed, the "evolution" of physics as a science is nothing but a continual increase in its generality. A century ago, for instance, phenomena of atomic spectra, of radioactivity, of chemical bonding, and many others were outside the ken of physics. This could be seen in at least two ways: either physics provided no way to even begin to treat such problems, or else, when it did, it gave blatantly wrong answers. From our present perspective, a century later, we can see that classical mechanics and classical thermodynamics were simply unable to cope with such problems. We can see now that the problems were *conceptual* ones and that theoretical physics a century ago was too restricted, too narrow, too special, to accommodate them, even in principle. At the time, of course, such problems were given no such interpretation; they were regarded as purely *technical* matters, requiring no modification of the underlying conceptual apparatus but only more cleverness in deploying that apparatus. In fact, it was authoritatively argued at just this time that physics was essentially complete as a conceptual discipline, that the primary remaining task for physicists was merely to measure its parameters with every greater accuracy. In short, just before physics was to undergo the most profound revolutions in its history, it was widely believed that it had already achieved its ideal state.

I have already indicated above how the two primary revolutions in physics in our century, relativity and quantum mechanics, came about. In each case, nineteenth-century physics had inadvertently restricted itself by means of two apparently harmless, and indeed quite invisible, hypotheses: that the velocity of light was infinite and that the quantum of action was zero. This was, of course, tantamount to requiring all velocities to be small, all energies low; nineteenth-century physics was really only the physics of material systems satisfying those conditions. In all other situations, it was forced either to stand mute or to be egregiously mistaken.

By contrast, the physics of the twentieth century (i.e., contemporary physics for us today) can accommodate arbitrary energies, arbitrary velocities. The generalizations that comprise modern physics have enormously extended the range of phenomena we can now accommodate. In the process, we have learned that the range of nineteenth-century physics consisted of what today are *limiting cases;* situations that are atypical, and *nongeneric,* in the context of the physics of today.

How could this happen? Only because, in the nineteenth century, material situations involving high energies and high velocities were *rare*. They could neither be produced technologically nor observed (or at least recognized) with the instruments then available. Because they were rare, and for that reason only, they could be discounted as technical anomalies, or given *ad hoc* "explanations" within the then-existing conceptual framework.

But we have seen this before (see section 1A above). Physics has discounted biology because organisms are *rare* in the class of material systems. On that basis, they have been presumed *special,* nongeneric *in the class.* But that is exactly the same kind of assumption that, as we have seen, characterized nineteenth-century physics; high-velocity, high-energy phenomena were *rare* in the nineteenth century, and hence, regarded as special and nongeneric. The truth in the latter case was quite the contrary of this; it is rather low-velocity, low-energy situations that are really from our present perspective the rare, atypical, nongeneric ones, however, prominent they may appear in everyday experience.

The generalizations that have allowed contemporary physics to encompass relativistic, and quantum-theoretic, phenomena have, of course, radically altered the face of physics itself; it is a far different thing today than it was a century ago. But in a certain sense, there is still a great deal of nineteenth-century physics, and seventeenth-century physics, that has survived these revolutions and generalizations intact. What other tacit presuppositions and limitations lie lurking there?

As I noted earlier, with respect to biological phenomena, contemporary physics is in exactly the same situation that nineteenth-century physics faced in the atomic and cosmological realms: it either stands mute or it gives the wrong answers. That is the simple fact. Once again, as in all similar situations in the past, the claim is that purely technical matters are involved and that the problem is simply one of *specializing* what already exists in an appropriate way. But history shows it to be at least equally likely that the problems are not technical but conceptual, that contemporary physics remains too special to accommodate the class of material systems we call organisms.

If so, it becomes a matter of finding, and removing, whatever tacit hypotheses are limiting the generality of contemporary physics in these directions. We cannot find them by looking at contemporary physics itself;

it, and everything in it, are already the consequences of imposing these very hypotheses. Rather we must retreat to an earlier conceptual stage. In the process, we shall find in fact that contemporary physics embodies a number of such restrictive hypotheses, and we shall see in detail the dramatic effects of removing them.

Chapter 3

Some Necessary
Epistemological Considerations

I FEEL IT is necessary to apologize in advance for what I am now going to discuss. No one likes to come down from the top of a tall building, from where vistas and panoramas are visible, and inspect a window-less basement. We know, intellectually, that there could be no panoramas without the basement, but emotionally, we feel no desire to look at it directly; indeed, we feel an aversion. Above all, there is no beauty; there are only dark corners and dampness and airlessness. It is sufficient to know that the building stands on it, that its supports, its pipes, and plumbing are in place and functioning.

3A. Back to Basics

Scientists, especially, are impatient with their basement of epistemology and ontology and what they call metaphysics. We are proud professionals who have reared not just a building but a temple, a monument. Our calling our genius, is in fact never to stop building but rather to push what we have received ever higher, to enlarge, to adorn, to move upward. Why take the time and the trouble to descend and contemplate anew uncongenial things that were settled long ago?

And so I must apologize for conducting the reader on a necessary trip back to the basement. It will only be for a short while, I promise, and I also promise that what we do there will be of importance.

Moreover, having been at the summit allows one to see the basement with new eyes. Many years ago, the mathematician Felix Klein wrote a series of books called *Elementary Mathematics from an Advanced Standpoint;* as the title suggests, the books constituted an illuminating return to the lower floors of mathematics, from the perspectives of the higher. And one should also ponder the reminiscence of Einstein:

> I sometimes ask myself, how did it come that I was the one to develop the Theory of Relativity? The reason, I think, is that a normal adult never stops to think about problems of space and time. These are things which he has thought of as a child. But my intellectual development was retarded, as a result of which I began to wonder about space and time only when I had already grown up. Naturally, I could go deeper into the problem than a child. . . .

Sufficient reason indeed to beat a strategic retreat. And that is what we shall do for the next several sections.

3B. The First Basic Dualism

Science is built on dualities. Indeed, every mode of discrimination creates one. But the most fundamental dualism, which all others presuppose, is of course the one a discriminator makes between self and everything else.

As Descartes argued long ago, the only absolute, undoubtable certainty lies here, and he put it with the ultimate terseness: *Cogito, ergo sum.* By *"cogito,"* Descartes meant the entire spectrum of the activity of his mind: perception, cognition, ideation, will, imagination. Also, although this gave him a great deal of trouble, because it involved something beyond this warm circle of certainty, he meant the capacity for action.

Thus, as Descartes says, we know our *selves*, without even having to look, by an immediate kind of direct apprehension and with a knowledge that brooks no skepticism.

Oddly, I have not been able to find a really good word that incorporates all of the activities of the self that we know with such immediate certainty. In physics, the word *observer* is often used, but as we shall see, this is too passive. There is the Freudian word *ego*, which is more encompassing than "observer" but has become ringed round with connotations irrelevant or misleading for our purposes. Perhaps best, as we have done, to continue to use the noncommittal word "self," though it seems rather drab and humble, and certainly insufficiently technical, for the exalted role that Descartes gives it.

At any rate, we know our self with ultimate certainty, even though this knowledge is *subjective;* it cannot be experienced as we experience it by anything else; at best it can only be reported. As noted, we encompass as belonging to the self, or contained within it, our perceptions, our thoughts, our ideas, our imaginings, our will, and the actions that spring from them. This is the *inner world*. Everything else is *outside*.

What else is there? Whatever it is, I shall call it the *ambience*. Most of us believe there are indeed many things in our ambience; this is the *external world*, the world of objective reality, the world of phenomena. That world is important to us, because our bodies are in that world, and to that extent at least, we must seriously care what goes on out there.

Much more could be, and has been, said about this fundamental dualism between the self and its ambience, but we shall need no more than the simple fact of its existence. Science, in fact, requires both; it requires an external, objective world of phenomena, and the internal, subjective world of the self, which perceives, organizes, acts, and understands. Indeed, science itself is a way (perhaps not the only way) of bringing the ambience *inside*, in an important sense, a way of importing the external world of phenomena into the internal, subjective world that we apprehend so directly. I shall have much more to say about this when we come to the idea of natural law, and especially, the idea of a *model* (see section 3F et seq. below). Indeed, as we shall see, the fact that inner, subjective models of objective phenomena *exist* connotes the most profound things about the self, about its ambience, and above all, the relations between them.

3C. The Second Basic Dualism

Our first basic dualism has separated the universe into a self and its ambience. For each of us, this separation is absolute, indubitable, and unequivocal, though it may be different for different selves. Our second basic dualism concerns the way we partition our ambiences, the way we *manage* our perceptions of the external world.

At this level, we have no universal principles to guide us, nothing *given* to us, like the distinction between the inner world of the self and the outer world, what we called the ambience. It rests rather on a consensus *imputed* to the ambience, rather than on some objective and directly perceptible property of the ambience. It is the dualism between *systems and their environments*.

Roughly speaking, a *system* in the ambience is a collection of percepts that seem to us to belong together. It would be hard to imagine a less precise definition of anything, but that is inherent in the very idea of system. The abstract concept of *systemhood* is indeed a very difficult one to grapple with, as is the related notion of *set-ness*. It is at the same time familiar in the concrete garb of everyday experience and alien when we attempt to characterize it in isolation, as a thing in itself, apart from any specific material embodiment.

Indeed, in mathematics, set-ness is such a basic and familiar notion that it took two thousand years for it to be recognized explicitly; even then, it took a strange mind (as contemporaries reported Cantor was) to see it and to deal with it. Once it was pointed out, and its central role in mathematical thought made explicit, then everyone saw it. Indeed, within a generation, and in the teeth of paradoxes it had already spawned, David Hilbert was saying, "From the *paradise* created for us by Cantor, let no man drive us forth."

The notion of system-hood is at that same level of generality and plays the same kind of role in our management of the ambience. As noted, it segregates things that "belong together" from those that do not, at least from the subjective perspective of a specific self, a specific observer. These things that belong together, and whatever else depends on them alone, are segregated into a single bag called *system;* whatever lies *outside,* like the complement of a set, constitutes *environment.*

The partition of ambience into system and environment, and even more, the imputation of that partition to the ambience itself as an inherent property thereof, is a basic though fateful step for science. For once the distinction is made, attention focuses on *system.* Systems and environment are thenceforth perceived in entirely different ways, represented and described in fundamentally different terms. To anticipate somewhat, system gets described by *states,* which are determined by observation; environment is characterized rather by its effects on system. Indeed, it is precisely at this point that, as we shall see, fundamental trouble begins to creep in; already here.

The growth of science, as a tool for dealing with the ambience, can be seen as a search for special classes of systems into which the ambience may be partitioned, such that (1) the systems in that special class are more directly apprehensible than others, and (2) everything in the ambience, any other way of partitioning it into systems, is generated by, or reducible to, what happens in that fundamental class. Newtonian mechanics, for instance, thought it had found such a class; so, today, does quantum theory. But it is, above all, a *special* class, embodying an equally special way of coping with the system-environment dualism itself. Whether this is enough is, at root, the basic question.

3D. Language

An essential part of the inner world of any self is one's language. It is a way, or reflects a way, of organizing percepts and perhaps even of generating them.

Language itself creates, or embodies, new dualism distinct from (but in many ways parallel to) those we have already discussed. Indeed, language is a unique and anomalous thing, whose acquisition, and even more, whose correct deployment, is a kind of miracle. I cannot dwell on these matters here but rather will concentrate on its essential role as an intermediary between the self and its ambience, and between one aspect or part of the self and another.

The first basic dualism inherent in language is that (1) it is a thing in itself and (2) permits, even requires, referents external to itself. These embody respectively what we will call the *syntactic* aspects of language and its *semantic* aspects. Roughly speaking, syntax pertains to what language *is,* as a thing in itself, while semantics pertain to extralinguistic referents. These referents may involve the self, or the ambience, or both, or even neither.

Let us consider syntactic aspects first. Syntax involves its own inherent dualism, which may be roughly described as the dualism between *proposition* and *production rules*. From a syntactical point of view, divorced from any external referents, propositions in the language are in general not *about* anything and are described entirely in terms of conventional symbol vehicles: letters, words, sentences, and so forth. The production rules are themselves propositions, but they do have referents, namely, other propositions *in the language*. Their role is essentially a dynamic one, to enable the construction of new propositions from given ones, or the analysis of given propositions into simpler ones.

The syntactical production rules of a language are its internal vehicles for what I shall call *inferential entailment*. The rules thus allow us to say, without consulting any external referent, that *one proposition, or group of propositions, implies others*. More generally, inferential entailment is a relation between propositions and means precisely that there is a string of production rules whose successive application will take us from some of them to the others.

Just as nobody has been able to characterize an organism in terms of a discrete list of properties, no one has been able to characterize a "natural language" (let us say English) in terms of a list of production rules. Indeed, if it were possible to do this, it would be tantamount to saying that a (natural) language can be completely characterized by syntactic properties *alone*, i.e., made independent of any semantic referents whatever. There have indeed been deadly serious attempts to do precisely this (see my remarks on formalization below). They have all failed, often rather dramatically, indicating (what might be obvious) that, in general, semantics cannot simply be replaced by more syntax. Nevertheless, the attempt to do so has served to extract various kinds of syntactical "sublanguages"; these will play an analogous rule, in the external world of the self, to the segregation of systems in one's external world or ambience. Indeed, as we shall soon see, there is more than just an analogy here.

We shall understand by a *formalism* any such "sublanguage" of a natural language, defined by syntactic qualities alone. That is, a formalism is a finite list of production rules, together with a generating family of propositions on which they can act, without any specification or consideration of extralinguistic referents. Thus, a formalism, as a fragment of natural language, *could* be "about" something (i.e., endowed with extralinguistic referents), but it *need not be*. A formalism, by its very nature, carries with it no "dictionary" associating its propositions with anything outside itself. It is propelled entirely by its own internal inferential structure, as embodied explicitly in its production rules. These and these alone determine the relations among the propositions of the formalism, which we have called inferential entailment.

As we shall see, the extraction of a formalism from a natural language has many of the properties of extracting a system from the ambience. Therefore, I shall henceforth refer to a formalism as a *formal system;* to distinguish formal systems from systems in the ambience or external world, I shall call the latter *natural systems*. The entire scientific enterprise, as I shall soon argue, is an attempt to capture natural systems within formal ones, or alternatively, to embody formal systems with external referents in such a way as to describe natural ones. That, indeed, is what is meant by *theory*.

A prominent trend, indeed a characteristic one, of contemporary science and mathematics is to try to dispense with extralinguistic referents entirely and replace them with purely syntactic structures that only recognize and manipulate the symbols of which the propositions themselves are built. This process is, naturally enough, called *formalization*. It involves the internalization of semantic referents, in the form of additional, purely symbolic, syntactic rules. It has never been better described than by S. C. Kleene (1950); I have quoted it often before, but it does no harm to repeat his words again here:

We are now about to undertake a program which makes a mathematical theory itself the object of exact mathematical study. . . .

The result of the mathematician's activities is embodied in propositions . . . we can contemplate the system of these propositions.

. . . As the first step, the propositions of the theory should be arranged deductively, some of them, from which the others are logically deducible, being specified as the axioms (or postulates).

This step will not be finished until all the properties of the undefined or technical terms which matter for the deduction of the theorems have been expressed by axioms. Then it should be possible to perform the deduction treating the technical terms as words *in themselves without meaning.* For to say that they have meanings necessary to the deduction of the theorems, other than what they derive from the axioms which govern them, amounts to saying that not all of their properties which matter for the deductions have been expressed by axioms. When the meanings of the technical terms are thus left out of account, we have arrived at the standpoint of formal axiomatics.

This idea of formalization, that the semantic aspects of language can *always* be effectively replaced by purely syntactic ones, will turn out to be another place where really serious trouble creeps in. Indeed, Gödel showed in effect that it was already false for Number Theory. It will turn out to be closely related to the reductionistic idea that there is always a "largest model," as I shall later describe in detail (see chapter 8). For the moment, however, I simply suggest the reader bear in mind the basic conclusion we can distill from the discussion above: *natural language is not a formalization.*

The study of formal systems is what comprises the subject of (in the broadest sense) *mathematics.* Its object is the universe of formal systems, just as real and significant a part of the self's internal world as are the natural systems one extracts from one's ambience. Seen in another way, mathematics is the study of inferential entailment, the art of extracting inferents from premises or hypotheses.

I conclude this brief consideration of language by pointing out two apects of natural language that will play key roles in what follows but that never end up as part of formalisms. These are (1) the use of the *interrogative,* to which I have already alluded, and (2) the use of the *imperative.* The latter, for example, is universally presupposed, even in mathematics; an algorithm, for example, is nothing but a strong of imperatives, ordering us to apply specific production rules to specific propositions, assuring us that *if* we do so, some definite end *will thereby be entailed.* In the world of natural systems, similar lists of imperatives constitute recipes, protocols, blue-prints, and the like, which govern *fabrication*. But, as will become apparent, the entailment process *embodied* by algorithms or recipes is very different than that governing their *application*. The difference, indeed, is precisely the difference between fabrication and physiology, which I contrasted earlier (see section 1C above). And the difference between them will provide another central feature of our overall enterprise.

3E. On Entailment in Formal Systems

Entailment is perhaps the central concept in the present work, as it is in the entire scientific enterprise and beyond. We have just met with one kind of entailment, namely, inferential entailment or implication within a formalism. As I stated, it constitutes a relation between propositions in a formalism; *P entails Q*, or *P implies Q*, if there is a string of production rules that take us from the proposition P to the proposition Q in the formalism. There are many ways to say this: P is a premise and Q is a conclusion; P is a hypothesis and Q is a theorem; etc. And of course, a formal system itself can be looked at as the totality of propositions Q entailed as theorems by the "axioms," using the system's inferential machinery or production rules in this fashion.

One of the endearing features of a formal system is that, *if it is consistent*, then truth necessarily percolates hereditarily from the postulates to the most remote theorem; inference can never take us out of a class of true propositions. By *consistent*, we mean only that proposition of the form $(P \vee \bar{P})$ are never theorems, where $\bar{P}$ is the negation of P (whatever P may be). Hence if the postulates are arbitrarily called "true," that same adjective may be safely applied to every inference or entailment drawn from them. In everything that follows, we always suppose we are dealing with formal systems that are consistent in the above sense. As we saw earlier, though, consistency is something we can seldom be absolutely sure of.

In any case, suppose that we step outside our formalism and contemplate one of its theorems P. That is, let us look at, or observe, our own formalism. From that perspective, we can *interrogate;* we can ask, for instance, what is it about that system which makes P a theorem? Put another way, we can ask: why is P true in the system? What, if anything, *entails* "the truth of P"?

To this question, several distinct kinds of answers can be given.

1. Obviously, the truth of P in a formal system depends on that system's axioms. For, by definition, P is entailed from precisely those axioms; this

is exactly what makes P true. So the axioms, which entail P, also play a role in entailing "the truth of P."

2. A different kind of role within the system is played by the production rules, the machinery that actually does the entailing. Different production rules, with the same axioms, might not entail P. Therefore, "the truth of P" is also partially entailed by the rules of inference that govern the system.

3. The actual entailment of P by the axioms, which constitutes the *proof* of the theorem P, involves more than just the inferential rules themselves. It is an explicit *list* of these rules, to be applied sequentially in a particular order. The first rule in the list is applied to the axioms themselves. The second rule is applied to the lemma arising therefrom. The third rule is applied to the result, and so on. The last rule on the list produces P itself. Such a list of production rules, each in an imperative mood, constitutes an *algorithm* or *program*. The exhibition of such an algorithm or list is another way of entailing "the truth of P" in our system.

These ways of answering the question "why is P true?" are clearly different and independent. Any of them can be changed or modified, independently of the others, and any such change may modify the status of "the truth of P." We can change an axiom, without touching the inferential rules or any algorithm drawn therefrom. We can change an inferential rule, without changing either the axioms or the list of which that constitutes our algorithm. Finally, we may change the algorithm, without affecting either the axioms or the rules themselves. And if we do make any one of these changes, thereby changing the status of "the truth of P" in the modified system, there is no guarantee that we can make further changes that restore "the truth of P" to its original status. Thus, for instance, if we change an axiom in such a way that P is no longer a theorem, we may not be able to make a corresponding change in the production rules, and/or the algorithm that constituted the original proof of P, so as to make P a theorem again. This is what we mean when we say the three ways we have answered the question "why is P true?" are independent.

Before going further, let us also note the obvious fact that the kinds of changes we have contemplated all come from *outside the formalism*. There is obviously no mechanism *within* the formalism for changing an axiom, or a production rule, or for applying a different algorithm. Furthermore, from the standpoint of the formalism, *anything* that happens outside is accordingly *unentailed*. To put it another way: A question like "why has this axiom been replaced by another?" can have *no answer at all* within the formalism itself. This is our first glimpse of a peculiar thing, which will later become of prime importance: namely, that though formal systems allow us to talk about entailment in a coherent way, from their standpoint everything important that affects them is itself *unentailed.*

So I have given three different kinds of answers to the question "why is P true?" The discussion I have provided should remind the reader of something we have seen before: namely, the Aristotelian discussion of the causal categories. Indeed, we have paralleled three of his four categories of causation; specifically, if we call the theorem P an *effect,* we may identify his idea of material cause of P with the axioms of a formalism, his idea of efficient cause of P with its production rules, and his idea of formal cause of P with the specification of a particular sequence or algorithm of production rules, generating a corresponding trajectory of propositions from axioms to P. At the moment, this should be regarded as no more than a curiosity; our main point is simply to indicate that *the Aristotelian analysis can be applied to any entailment structure,* simply by (as he did) asking "why?" about it.

The reader may not be surprised to note that we do not see a formal analog of Aristotle's fourth causal category, which he held to be the most significant; namely, the category of final cause. Perhaps the reader would be more surprised to see finality asserted; for centuries past, it has been part of the essential core of science itself that science and finality are incompatible. I shall discuss this matter, which is in fact one of the key inheritances we receive from Newtonian mechanics, when we come to consider natural systems (see sections 4I, 5K below). But already at this point, in the internal world of formalism, we can say a few important things about finality.

In what follows, *telos* is not involved; as we shall soon see, finality and teleology are in fact very different things. In completely formal terms, we may note that final causation appears anomalous, when compared with the other categories of causation. Formally, to say that something is a final cause of P is *to require P itself to entail something;* in every other case, to say that something is a cause of P means only that it *entails P.* Final cause thus requires something of its effect P; in all other cases, nothing is required of P beyond the passive fact of its entailment.

Moreover, in addition to requiring its effect P to entail something, a final cause of P must entail *the entailment of P itself.* It is this peculiar reflexive character of final causation, visible here in purely formal terms, that is primarily responsible for its anomalous position.

Let us recast these considerations in another way, since they will become crucial for us later. In any formalism, there is a kind of natural flow from axioms to theorems, very much like the familiar unidirectional flow of time. Indeed, the formal analog of "time" is embodied in the idea of *sequence,* the order of application of production rules or inferential operations in proofs and algorithms. This flow of "formal time" is irreversible, just as real time is, and as we shall see, for exactly the same reasons. In it, the axioms are always *earlier* than any of their consequents; a proposition P is *later* than another Q if it is implied by it, if there is a proof of P with Q as hypothesis.

The three "traditional" causal categories (formal, material, and efficient causation) always respect this flow of "formal time", in the sense that "cause" Q always precedes effect P. Final causation gives the *appearance*, at any rate, of violating this flow, in the sense that the effect of P seems to be acting back on the causal process that is generating it; it appears that the "future" is actively affecting the "past." I say "appears" because this (traditional) interpretation of finality confuses P with its final cause; it is not the effect P, but the final cause of P, that must operate on the process by which P is generated. The temporal anomaly remains, however; final cause clearly cannot fit *within the same temporal sequence* in which the other causal categories harmoniously operate.

The rejection of finality in science is usually cast in this temporal context, in the form of an unspoken "Zero[th] Commandment" permeating all of theoretical science: "Thou shalt not allow the future to affect the present."

The upshot of this discussion of finality is the following: in purely formal terms, a concept of final causation requires modes of entailment that are simply not generally present in formalisms. In a typical formalism, a proposition P entails only its consequents under the given inferential rules. Further, there is generally nothing in a formalism that *entails an entailment*. Both are required to make a concept of finality formally meaningful; formalisms generally contain neither. There is, however, nothing inherently impossible about them, and for the moment, I simply suggest keeping an open mind about formalisms *rich enough in entailment* to admit a meaningful category of final causation, at least in the limited sense I have described. In effect, I am suggesting, on formal grounds, the possibility of *separating finality from teleology*, of retaining the former while, if we wish, discarding the latter. Later (see chapter 5), I will argue that the incorporation of finality into our scheme of things, in the form of the additional modes of entailment it requires, is not only possible, it is crucial.

3F. On the Comparison of Formalisms

I have argued above that mathematics, in the broadcast sense, is the study of formalisms and that formalisms, in their turn, are parts of natural language whose inferential or entailment structures are defined in purely syntactic terms. But mathematics comprises more than this. For one thing, mathematics selects, from the plethora of formalisms available, only a rather small number of protracted scrutiny; the selection process, which gives mathematics its form as a human activity, is not itself part of any formalism. For another thing, once a variety of formalisms is so selected, the formalism themselves become elements in a potential mathematical universe.

More precisely, formalisms may be *compared,* in terms of their respective inferential or entailment structures. When are two formalisms, which *look* different in terms of their axioms and production rules really "the same formalism" in terms of the body of propositions (theorems) they generate? When does one formalism subsume another, so that the second can be in some sense generated from the first, or embedded in it? And above all, is the machinery for dealing with such questions, i.e., with the comparison of formalisms, itself a formalism? In the present section, I shall briefly describe some of these questions, which will arise again in other contexts later; it turns out to be instructive to look at them here, in a purely formal context, free of the murky epistemological embroidery that obscures the fundamental issues elsewhere.

The problem of comparing formalisms is at root one of classification or taxonomy; it does for mathematics what Linnaeus did for biology. The basic underlying idea for such comparisons goes back to Euclid and is embodied in the idea of *similarity* or *congruence;* related geometric figures can be brought into coincidence by applying some kind of transformation to the underlying space. How related the figures are is measured in some sense by the complication of the transformation required to make them coincide. For instance, Euclidean congruence requires only rigid motions (rotations, translations, reflections); Euclidean similarity (equal angles) allows transformations of a more *general* character, and so on until we get to topology (in which figures are called congruent if they is simply a continuous map that brings them into coincidence).

This kind of taxonomic classification permeates mathematics (and much more beyond); the general study of similarity analysis in fact involves one of the archetypal aspects of human thought. It would take us too far afield to consider it in detail here (see *AS*), but I offer one more example from mathematics itself.

Let S be a linear vector space, a set of things called vectors. These vectors may be added and may be multiplied by scalars (numbers), subject to the normal rules governing these operations, which everyone knows. Inserting a coordinate system into S allows us to express any vector as an array of numbers and converts vector addition and scalar multiplication to the familiar manipulation of these arrays. Different coordinate systems will clearly associate each vector with a different array of numbers.

A *linear transformation* T of one such vector space S_1 into another S_2 is an ordinary mapping of the set of vectors S_1 into the set of vectors S_2, which respects the vector operations; it maps sums to sums, and scalar multiples to scalar multiples, in the familiar way. As such, it is already an instrument for *comparing* S_1 and S_2, but that is not our present concern.

If S_1 and S_2 come along with a particular coordinate system, then T itself translates into a particular array of numbers, a *matrix*. The same transformation T will give rise to different arrays, different matrices, in different coordinate systems. Thus, just looking at matrices, i.e., at arrays of numbers, we can ask when two such arrays come from the same linear transformation T, only seen in different coordinate systems. This clearly establishes a relation on *matrices* (similarity); two matrices are similar if they "differ by coordinate transformations." Specifically, if A and B are matrices, then they are similar if and only if there is a coordinate transformation g_1 of S_1, and a coordinate transformation g_2 of S_2 such that, for every vector v in S_1

$$g_2 A(v) = B g_1(v)$$

or, in more familiar abstract terms, if

$$g_2 A\, g_1^{-1} = B.$$

This kind of relation can be embodied in a diagram that exhibits the mappings involved, namely,

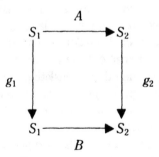

The similarity condition requires this diagram of mappings to *commute;* the two paths going from upper left to lower right in the diagram must always agree.

I have taken this digression, first, to show that mathematical objects besides geometrical figures can be compared (i.e., congruence is an idea of more general currency than just geometry), and second, to exhibit in a familiar situation the kind of archetypal commutative diagram that is the hallmark of comparison in mathematics.

Now let us return to the man question, namely, the comparison of *formalisms*. What there is to be compared here, i.e., to be brought into congruence, to the extent possible, is the inferential structures that characterize the formalisms. The result, if we are successful, will be a commutative diagram of the type we have just seen.

So let us say we have a *formalism* F_1, and another formalism F_2 we wish to compare it with. Each of these formalisms possesses, of course, its own inferential structure, its own set of production rules and axiomatic propositions on which they act to generate the consequents or theorems that constitute the respective systems. I will indicate these autonomous inferential structures in the schematic way in figure 3F.1.

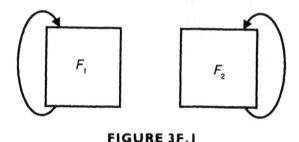

FIGURE 3F.1

So far, F_1 and F_2 are only, so to speak, talking to themselves. In order to compare them, we need to get them to talk to each other, or better, to express what each formalism says to itself in the language of the other. In other words, as with comparing any two languages, we need to make a *dictionary*, in fact, a pair of dictionaries. The first of them, which will translate from F_1 to F_2, I shall call the *encoding dictionary*, or simply the *encoding* (of F_1 into F_2). The other, translating from F_2 back to F_1, I shall call the *decoding dictionary*, or just *decoding*. I note explicitly that, in general, these dictionaries are not just inverses of each other; we do not require any relation at all between them.

So far, these encodings and decodings are just set-theoretic maps from propositions in one system to propsoitions in the other, which mandate a *synonymy*. Putting all the arrows together, we have a diagram (figure 3F.2) of the form which is just figure 3F.1 augmented by the dictionaries.

So far, the dictionaries are arbitrary. As such, there is no reason why they should respect the inferential structures in F_1 and F_2. More explicitly, given a proposition P in F_1 the diagram gives us two ways to deal with it:

DECODING

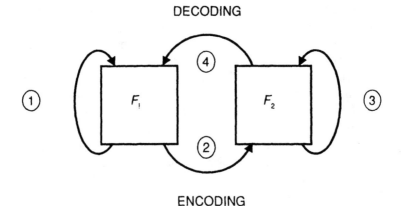

ENCODING

FIGURE 3F.2

1. We can apply to P the inferential machinery in F_1 itself and look at the resultant propositions entailed thereby.
2. We can *encode* P (if it is indeed allowed by the encoding dictionary we are using) into a proposition P' in F_2. Then, using the inferential machinery of F_2, we can look at the propositions in F_2 entailed by P'. And finally (again, if it is allowed by the decoding dictionary), we can decode these back into propositions in F_1.

Using general dictionaries for encoding and decoding (i.e., using an unconstrained specifications of synonymy), there is no reason why these two procedures should agree or coincide. That is, using the numberings of figure 3F.2, we will generally find that

$$1 \neq 2 + 3 + 4$$

Thus, the diagram *does not commute*. In such a case, our choice of encoding and decoding establishes no relation between the inferential structures of F_1 and F_2.

It may, however, be that we can *find* encodings and decodings for which the diagram of figure 3F.2 *always commutes.* in such a case, we have in fact brought at least a part of the inferential machinery of F_1 into congruence with a corresponding part of the inferential machinery of F_2. We will then say that F_2 is a *model* of F_1, or equivalently, that F_1 is a *realization* of F_2. We will also say that a *modeling relation* exists between the two inferential structures.

We should note parenthetically here that the word *model* is overworked and has been used in a whole host of different, sometimes unrelated, and even contradictory senses. There is, for example, a well-developed *Theory of Models*, employed mainly to study the consistency of axiom systems in Foundational Studies in mathematics. The use of the term "model" in this context is not quite the same as mine; in fact, it is more closely akin to what I have called "realization." However, since I will never use the word "model" in any other sense than the one I have specified, there will be no danger of *internal* confusion. I simply want to warn the reader at this point that the danger of equivocation on "model" is unusually great.

A host of truly marvelous things follows from the establishment of a modeling relation between formalisms (see *AS*). I shall describe some of these later, in a more appropriate context. The main thing about it, of course, is that we can use the inferential structure of the model to study that of its realization, to *predict* in effect, from an encoded hypothesis P' (via the pathway $2+3+4$ in the diagram) theorems of F_1 from theorems in F_2.

An enormous amount of this kind of formal modeling of one kind of inferential structure in another occurs in mathematics. Indeed, whole fields (algebraic topology, for example, see section 5I below) consist of nothing else. I have in *AS* described and illustrated these activities at great length and there is no need to repeat it here. I shall rather content myself with pointing out some of its general features, which will be of importance to us later.

The first matter of importance is to note that, from the standpoint of the formalisms being compared, *the encoding and decoding arrows in figure 3F.2 are unentailed.* In fact, they belong to neither formalism, and hence, *cannot* be entailed by anything in the formalisms. The comparison of two inferential structures, like F_1 and F_2, thus inherently involves something outside the formalisms, in effect, a *creative act,* resulting in a new kind of formal object, namely, the modeling relation itself. It involves *art.*

The second matter concerns whether this creative act can itself be formalized, i.e., whether the study of comparison of formalisms is itself a formalism. In a nutshell, the answer is *yes, in a sense.* The name of that formalism is the *Theory of Categories;* the qualification is that Category Theory, like Number Theory, like Set Theory, or like natural languages themselves, cannot be *formalized,* in the sense of Kleene quoted previously. Indeed, many mathematicians have wondered aloud, over the years, whether Category Theory is even a part of mathematics.

However, Category Theory comprises in fact the general theory of formal modeling, the comparison of different modes of inferential or entailment structures. Moreover, it is a stratified or hierarchical structure, without limit. The lowest level, which is familiarly understood by Category

Theory, is, as I have said, a comparison of different kinds of entailment in different formalisms. The next level is, roughly, the comparison of comparisons. The next level is the comparison of these, and so on.

The final matter I wish to draw attention to here is the following. In a precise sense, any formalism F has a "biggest" formal model, namely, F itself. But as we have seen, even in mathematics we find entailment structures that are not formalizable. Let us use natural language for the sake of illustration. There is certainly a lot of entailment in natural language; suppose we want to model it. That is, suppose we want to place a natural language into the box occupied by F_1 in figure 3F.2 above. Then what?

In fact, we can find a host, an *unlimited number, of distinct formalisms* F_2 that we can put into a modeling relation with the language. No one of them, nor indeed, no aggregate of them, can replace the language, in the sense of completely duplicating its entailment structure. In short, the totality of formal models of something that is not itself a formalism to begin with is

1. indefinitely large, and
2. is not itself a formalism.

These will turn out to be pregnant and profound conclusions. We can get some sense of their ultimate import if we replace the word "formalism" with the word "machine." Much lies ahead of us before we can make this kind of substitution sensible. But it will turn out that the conclusions we have drawn above, which have so far concerned only those formal entailment structures specifiable in terms of syntax alone, can be exported to comparison of *any* kind of entailment structures whatever. I now turn to the question of whether there are any others, and if so, what they might be like.

3G. Entailment in the Ambience: Causality

I now turn from the internal, formal world of the self to the external world that constitutes its ambience, the world we have come to look upon as populated by *natural systems* and their environments. I thus turn to the world of science, in the broadest sense.

The fundamental question for us, at this point, is the following: is there, in this external world, any kind of *entailment,* analogous to the inferential entailment we have seen between propositions in a language or formalism? Obviously, if there is not, we can all go home; science is not only impossible but also inconceivable.

This kind of question has always been difficult, because we come by our knowledge of the ambience at second hand. As philosophers have pointed out for millennia, all we perceive directly are our selves, together with sensations and impressions that we normally interpret as coming from "outside" (i.e., from the ambience), and that we merely *impute*, as properties and predicates, to things in that ambience. The things themselves, the *noumena*, as Kant calls them, are inherently unknowable except through the perceptions they elicit in us; what we observe are *phenomena*, which are to an equally unknowable extent corrupted by our perceptual apparatus itself (which of course also sits partly in the ambience).

We can simplify things somewhat if we ask the more restricted question: is there any kind of entailment at the level of phenomena? Or, stated otherwise: does it *appear* to us that a phenomenon can entail another? The problem is still difficult, because entailment at this level is a *relation* between phenomena (just as inferential entailment is a relation between propositions), and we usually do not directly perceive relations. Indeed, a relation between phenomena depends on a double imputation: the first from sensation to phenomena, the second from phenomena to relations between them. Thus, if our knowledge of phenomena is already once removed from the ambience, any talk of entailment, or any other kind of relation between phenomena, is twice removed. On top of all of this is a further problem, that what we *do* perceive is only a sample of what we *could* perceive and the problems of induction arising therefrom; see section 2C above.

It goes without saying that most of us can adduce the most compelling, convincing subjective evidence for believing that, and acting as if, there are indeed entailment relations between phenomena. But the question is rife for rampant skepticism; despite the combined efforts of countless philosophers, there is no way to *entail* the existence of such relations from anything else (i.e., from anything in the internal world of the self, or anything that the self draws from, or imputes to, the ambience). To such a skeptic, indeed, there is little to distinguish science from paranoia (which is basically a search for, or a belief in, entailments that are in some sense not there).

Nevertheless, it is hard to believe, for instance, that we could use natural language, in its semantic role of bringing external referents inside, if there were not a great many phenomenal entailments; semantic language by its very nature imputes hordes of entailments to the ambience, without going really dramatically astray. For this, and similar (albeit subjective) reasons, we will suppose that relations of entailment do indeed exist between phenomena; the question then becomes not whether, but when, such relations hold.

It was, of course, Aristotle who associated the notion of entailment between phenomena with the question "why?" and answered it with a "because." Indeed, the pair consisting of the question "why *A?*" and the answer "because *B*" precisely asserts an entailment of *A* by *B,* and hence, an *explanation* of *B* in terms of *A.* In this way, entailment relations between phenomena are subsumed under the general framework of *causality.* To the extent that science is the study of entailment relations between phenomena, Aristotle correctly identified science with the study of "the why of things" and scientific explanation with the elucidation of causal sequences.

Historically, Aristotle elaborated his view of the causal categories in terms of human artifacts (i.e., statues, goblets, houses) rather than in terms of animate or inanimate nature or in terms of formalisms. Nevertheless, as we have seen, his analysis holds good wherever there are relations of entailment *of any kind,* even in the world of formal systems, where entailment means inference. Accordingly, his analysis also applies to the world of natural systems that populate the ambience; as we shall see abundantly later, it permeates the whole of contemporary science, though in such a shrunken and distorted form that it takes a special effort of retrieval to make it manifest.

We shall thus accept this view, that entailment relations can exist between phenomena and that their study comprises causality; hence science and causality are to that extent synonymous.

I turn now to the last of our preliminary considerations, namely, the establishment of relations *between* the two entirely different kinds of entailment we have been considering. I have talked about *inferential entailment* in internally generated formalisms, governed by inferential rules that generate new propositions from given ones. And I have talked about *causal entailment,* relating phenomena arising in the ambience or external world. My final task is to show that these two entirely different modes of entailment are themselves related. The assertion of this relation is embodied in the concept of Natural Law; the crucial instrument in establishing the relation is the concept of *model.*

3H. The Modeling Relation and Natural Law

Over the preceding several sections, we have been mainly concerned with the concept of entailment. We have in fact found two entirely different realms in which entailment is meaningful. First, in the interior world of the self, I have called attention to inferential entailment and embodied it in the inferential structure of formalisms or formal systems. Second, in the outer world of the ambience, I identified a different kind of entailment, entailment between phenomena. As I said, this is the province of causality.

There are many parallels between these realms of entailment, which I am, in fact, presently in the process of making explicit. We have already seen, for example, that the Aristotelian analysis that led to his ideas about causation in the ambience actually apply to any realm possessing a notion of entailment; thus, in particular, we saw (see section 3E above) that his analysis was equally meaningful for syntactic entailment in formal systems. We also saw, in the preceding section, that the modes of entailment manifested by different formal systems may be compared; the vehicle for such comparison of entailment in different formal systems was the establishment of a modeling relation between them.

In the present section, we shall see that the concept of a model provides in fact a general method for comparing entailment structures of *any* kind. Just as the Aristotelian analysis may be applied to any mode of entailment, so too can modeling relations be established, *mutatis mutandis,* between entailment structures of arbitrary kinds. I shall now indicate in particular how we can compare inferential entailment in a formal system with causal entailments, relating a bundle of phenomena that we extract from our ambience and identify as a natural system.

A modeling relation between causal entailment in a natural system and syntactic entailment in a formal one provides a concrete embodiment of the concept of *Natural Law.* It is worth spending a moment discussing Natural Law, for it provides the explicit underpinning on which all of science rests.

Natural Law makes two separate assertions about the self and its ambience:

1. The succession of events or phenomena that we perceive in the ambience is not entirely arbitrary or whimsical; there are relations (e.g., causal relations) manifest in the world of phenomena.
2. The relations between phenomena that we have just posited are, at least in part, capable of being perceived and grasped by the human mind, i.e., by the cognitive self.

Science depends in equal parts on these two separate prongs of Natural Law. The first, which says something about the ambience, asserts that it is in some sense orderly enough to manifest relations or laws. Clearly, if this is not so, there can be no science, also no natural language, and most likely, no sanity either. So it is, for most of us at any rate, not too great an exercise of faith to believe this.

The second part of Natural Law says something about ourselves. It asserts that the orderliness of the ambience is (to some unspecified extent)

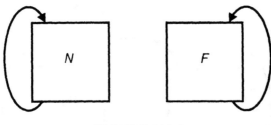

FIGURE 3H.1

discernible to, and even more, is articulable by, the self. It asserts then that the posited orderliness in the ambience can be matched by, or put into correspondence with, some equivalent orderliness within the self.

In other words, the first part of Natural Law is what permits science to exist in the abstract. The second part of Natural Law is what allows scientists to exist. Clearly, concrete science requires both.

I am now going to show how the modeling relation provides an explicit embodiment of Natural Law. Specifically, the causal entailments manifested by a natural system provide the orderliness required of the ambience. Inferential entailment in a formal system is a way of providing the orderliness required of the self. The art of bringing the two into correspondence, through the establishment of a definite modeling relation between them, is tha articulation of the former within the latter; it is in effect science itself.

My discussion will exactly parallel the one given earlier. Thus, let us suppose we are given a natural system N, and a formal system F (figure 3H.1). Just as before, the two arrows schematically represent the respective entailment structures; inference in the formalism F, causality in N. And as before, a comparison of these different kinds of entailment structures requires the establishment of *dictionaries*, one for encoding the phenomena of N into the propositions of F and another for decoding from propositions of F back to phenomena in N. Such dictionaries give us exactly the same diagram we have seen before (figure 3H.2).

The concept of such a dictionary is, of course, precisely what endows a syntactically defined formalism with external referents. In this light, the commutativity of a diagram like that of figure 3H.2, and the resultant modeling relation it embodies, is just an expression of the possibility of using *syntactic* truth and *semantic* truth consistently. As I have said several times before, natural language as we know it could not otherwise exist.

Intuitively, the encoding arrow ② is associated with the notion of *measurement*. In physics, for example, a measurement process is precisely geared to associate a *number* with an event or phenomenon in N. A number is an abstract object, a mathematical entity. Thus, a meter serves basically

DECODING

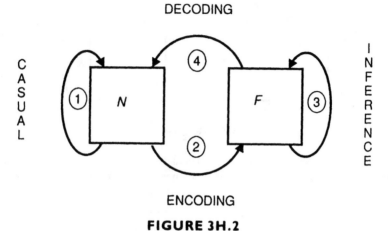

ENCODING

FIGURE 3H.2

as a transducer, associating numbers with phenomena. It is in fact one of the basic beliefs of physics, made quite explicit in quantum theory, that every observation, i.e., every material interaction of the self with its ambience, can be equivalently expressed in terms of an appropriate family of numerical measurements, but that is not really important to us now. Indeed, it may well be false; it does not change the argument.

It is not perhaps generally appreciated, especially by experimentalists (i.e., by those who actually perform measurements) that any measurement, however comprehensive, is an act of *abstraction*, an act of replacing the thing measured (e.g., the natural system N) by a limited set of numbers. Indeed, there can be no greater act of abstraction than the collapsing of a phenomenon in N down to a single number, the result of a single measurement. From this standpoint, it is ironic indeed that a mere observer regards oneself as being in direct contact with reality and that it is "theoretical science" alone that deals with abstractions.

In any event, the decoding arrow ④ in figure 3H.2 represents a *de-abstraction*, the association of a phenomenon in N with a proposition in F. It is thus a kind of "inverse" measurement, going from propositions to events.

Just as before, we note that there are two separate paths in the diagram: namely,

$$① \text{ and } ② + ③ + ④.$$

Each of these paths takes us from phenomena in N to phenomena in N. The first of them (the path ①) represents causal entailment within N; it is essentially what an observer, who simply sits and watches what happens in

N, will see. The second path, however, involves more. First we must encode, via the arrow ②, from phenomena in *N* to propositions in *F*. Next, we must use these propositions as *hypotheses*, on which the inferential machinery of the formal system *F* may operate. That machinery, the entailment structure within *F*, is what we have denoted by the arrow ③; it generates theorems in *F*, entailed precisely by the encoded hypotheses. The final step is, to the extent permitted, to *decode* these theorems back to the phenomena of *N*, via the arrow ④. At this point, the theorems we have thus generated become *predictions* about *N*.

The formal system *F* is then called a *model* of the natural system *N* if we always get the same answer, whether we follow path ①, or whether we follow the path ② + ③ + ④ As before, the establishment of a modeling relation between *N* and *F* serves to bring their respective entailment structures into at least a partial coincidence. To that extent, then, we can learn about one by looking at the other. And to that extent, modeling relations are nothing more than embodiments, in concrete situations, of natural law as I discussed it above.

As I have described it, the modeling process compares causal entailment in *N* with inferential entailment in *F*; if we are successful in establishing such a relation, then *F* is the model; *N* is a *realization* of that model. But it is essential to note that the roles of *N* and *F* can be interchanged. That is, instead of starting with a natural system *N*, and looking in effect for a formalism *F* that models it, we could start with a formal system *F* and ask for a natural system *N* whose causal entailment provides a model for inferential entailment in *F*. This, it will be recognized, is not simply an interchange of the arrows ② and ④ in figure 3H.2 above. It constitutes what I shall call *the realization problem*. At this level, it appears innocent enough; as we shall see, however, its consideration involves modes of entailment falling completely outside contemporary science. Indeed, the entailments required to deal with it are closely related to those that characterize *finality*, as I have described above. Ultimately, it will turn out that the absence of precisely such modes of entailment from our inherited scientific arsenal is what makes biology, and especially the origin of life, so hard.

We can already see some peculiar epistemological issues inherent in the diagram of figure 3H.2 that did not arise earlier. Specifically: what is the status of the encoding and decoding arrows in that diagram? We already saw, in the exactly similar diagram of figure 3F.2, that the encoding and decoding arrows were themselves *unentailed*. But at least they could themselves be considered as formal objects, since at that point we were comparing syntactic entailment in two formalisms. But now, we are com-

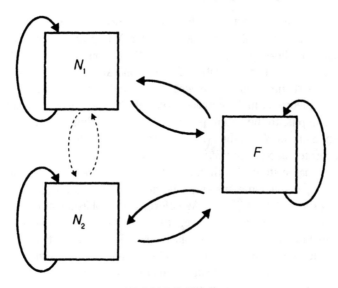

FIGURE 3H.3

paring syntactic entailment in a formalism with causal entailment in a natural system. The encoding and decoding arrows in this case are still *unentailed*, but it is no longer clear *how* they could be entailed, or from what. These arrows are not part of the natural system *N*, nor even of its environment; although they pertain to the ambience, they do not belong to it. Neither do they belong to the formal world of the self either; they look like mappings, but they do not compare formal objects; hence they cannot be mappings in any formal sense. Thus these arrows, which play the central role in comparing causal and inferential entailment, and hence, in the operation of Natural Law itself, turn out to possess a new and ambiguous status, equally within, and outside of, both the self and its ambience.

Modeling relations between natural systems and formal systems are wondrous things in many ways. Indeed, the innocuous-appearing diagram of figure 3H.2 has many remarkable ramifications. I have discussed many of these at great length elsewhere (see *AS*) and thus need not repeat most of it here. I can, however, use the concept of a formal model to complete my discussion of the comparison of entailment structures by showing how it can be used to compare causal entailments between two *natural* systems, say N_1 and N_2. We can do this by contemplating the diagram shown in figure 3H.3.

In this case, the two natural systems N_1, N_2 *realize a common formal system F*. It is clear that we can use the respective encoding and decoding arrows from N_1 and N_2 to *F* to construct the dotted arrows in the diagram,

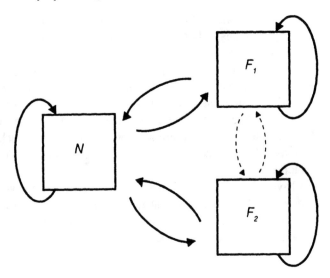

FIGURE 3H.4

in effect, to construct a dictionary from phenomena in N_1 to phenomena in N_2. The commutativities required by the assertion of modeling relations between N_1 and F, and N_2 and F, ensure that the whole diagram commutes, in the obvious sense. Hence N_1 and N_2 are *models of each other;* their respective structures of causal entailment are, to that extent, brought into congruence through the intermediary of a common formalism F, whch they separately realize.

This modeling relation between two natural systems N_1, N_2 is of the most profound importance; I shall call it *analogy.* It means that we can learn about one natural system N_1 by looking at a different, perhaps completely different, natural system N_2. This is familiar enough; under the rubric of *scale modeling,* engineers have long exploited the relation of analogy. In physics, we have already (see section 2C above) alluded to the "Mechano-Optical Analogy" of Hamilton and the wonders it spawned. Still more dramatic, perhaps, are the vistas opened by analogy for the further elucidation of causal entailment in natural systems that are, in conventional physical terms, of entirely different structures: organisms and "machines", organisms and social systems. This is another way of seeing, what I alluded to earlier, that reduction to a common set of material constituents is not the only way, nor even a good way, of comparing natural systems.

The complementary diagram to that of figure 3H.3 is exhibited in figure 3H.4. Roughly, it is the situation in which a single natural system N possesses two formal models.

Once again, we can construct the dotted arrows in the diagram from the posited modeling relations between N and F_1, and N and F_2, and thus compare syntactic entailment in F_1 and F_2. Clearly, the *formal* relationships between F_1 and F_2 arising in this way bear immediately on the problem of reductionism. For instance, we can ask, for *the class of all formalisms F that model N*, what is the formal structure of such a class? Is there a largest model in the class? Or are there causal entailment structures (i.e., natural systems N) in the ambience, as there are within the self, for which there is no largest model, for which the class of all formal models does not determine a formal model?

Thus we have come full circle, back to the problems with which we started. And with this, we are nearly at the end of our tour of the epistemological basement of contemporary science. In a moment we shall turn to a contemplation of what can be built thereon.

31. Metaphor

As we have seen, the modeling relation is intimately tied up with the notion of prediction. Natural Law, as embodied in modeling relations, thus equips us to look into the future of things; insofar as the future is entailed by the present, and insofar as the entailment structure itself is captured in a congruent model, we can actually, in a sense, pull the future of our natural system into the present. The benevolence of Natural Law lies in assuring us that such miracles are open to us, but it does not extend to telling us how to accomplish them; it is for us to discover the keys, the encodings and decodings, by which they can be brought to pass.

Thus, if we want to predict, to tell the future, there is much work to be done in the present. But most of us are lazy; we would like to decouple benefits from costs, to the fullest extent possible. A large part of the cost imposed by Natural Law, in return for the benefit of prediction, lies in finding the right encodings. But to what extent do we really need these encodings? Perhaps we can presume a little on Natural Law and get away without them.

At first sight, this does not look feasible; if we want to predict something about a particular natural system, it seems evident that we must specify that system, i.e., that we must encode it somehow. But on the other hand, prediction itself involves nothing more than *decoding* from a model or formalism. Certainly, if we already have a model, we can forget about how our natural system was encoded and obtain correct predictions just by decoding from the formalism. Perhaps this situation has wider currency;

perhaps we can decode something about the future of some natural system without explicitly encoding its present at all.

If we can do this, then Natural Law provides us to that extent with a nonspecific crystal ball whereby we seem to derive its benefits with only half the work. This is the essense of *metaphor:* decoding without encoding, in a sense, only the top half of our modeling relation.

All science, and biology in particular, is replete with such metaphors. They have in fact been of profound importance in the history of science and in many areas continue to play a major role; in fact, they constitute what there is of theory in these areas. Let us accordingly look briefly at a few of them.

Perhaps the most important for our purposes is the machine metaphor of Descartes, to which I have already alluded (see section 1C above), and about which I shall have much more to say (see chapter 7 et seq.). It asserts that things about machines can be decoded into predictions about organisms, without the benefit of any specific encodings going the other way. Machines thus become our crystal ball, our one-way mirror for looking into the organic world, without needing to look out again. We do not need to dwell further on the crucial role this metaphor continues to play in shaping the outlook of biology.

Another one of enormous current importance, which looks different from the Cartesian metaphor, but which is in fact closely related, is what may be called the open system metaphor. This is of relatively recent origin, at least in its present form; it was most explicitly articulated by Ludwig von Bertalanffy in the mid-1930s. Bertalanffy drew attention in particular to the metaphorical relation between what happens in the vicinity of stable point attractors (stable steady states) of open systems and the empirical facts of embryonic development: pattern generation or morphogenesis. In this metaphor, we seek to decode from the former into the details of the latter, again without the benefit of any specific encodings going the other way. It was this general metaphor, embodied in particular submetaphors by Rashevsky and Turing, that sent physicists like Prigogine scrambling to modify thermodynamics to accommodate them; see *AS*. In a somewhat different direction lie the more formal metaphors of Catastrophe Theory, first proposed by Thom. There are nowadays many variants of this basic metaphor, all being pursued with great diligence.

At root, such metaphors are pursued in the belief, or expectation, that they can in fact be turned into models. Thus, for example, the expectation is that any particular natural system, like a developing frog embryo, can be explicitly encoded into *some* formal open system, in such a way that a modeling relation is thereby established, or, more generally, that any organism can be explicitly encoded into some machine in such a way as to complete the modeling diagram so that commutativity holds. Meanwhile, the metaphor itself allows us to derive many of the benefits of the modeling relation even in the absence of these encodings.

To proceed metaphorically in the above sense is, of course, not an unreasonable thing to do. It is also clear, however, why experimentalists find such metaphors troubling and why they occupy an anomalous position in what passes nowadays for philosophy of science. For by giving up encoding, we also give up *verifiability* in any precise sense. Thus, experimentalists interested specifically in, say, a developing sea urchin, derive no tangible help from a metaphor. They need something to verify, couched in terms of some specific observation, or experiment, that they can perform. That is to say, they need precisely what is missing in the metaphor; they need the encodings. Hence the general indifference, if not active hostility, manifested by empiricists to theory couched in metaphorical terms. Metaphor is indeed immune to such verification; insofar as science is identified with verification, as it is currently fashionable to do, metaphor is not even science. Nevertheless, it is clear that metaphor can embody a great deal of truth. And as with all crystal balls, it does have the irresistible attraction of offering something for free.

Metaphor exists on the purely formal side as well. In the Theory of Categories, for example, it manifests itself in the concept of functor (see section 5I below). The turning of metaphor into model, in those terms, is expressed in the concept of natural transformation. The whole idea of the Theory of Categories arose initially in this way, and it is illuminating to continue to regard it in this light. In some sense, it is precisely the unique metaphorical aspects of Category Theory that generate qualms in many mathematicians regarding it, which run quite parallel to those of any empiricist.

Chapter 4

The Concept of State

AS WE have seen, the ambience or external world is traditionally regarded as being composed of systems (natural systems) and their environments. In the present section, we will begin to look into this matter more deeply, using the concepts we have already developed.

4A. Systems and States

Central to the notion of natural system is the attendant notion of *state*. As I remarked earlier, and as we shall see in more detail in the present chapter, *system* and *state* have become essentially coextensive; systems are described in terms of their possible states, while their environments are not (and indeed, cannot be).

As is true with all the deep concepts of contemporary science, the idea expressing systems in terms of states, and everything that happens in systems in terms of state transitions, goes back to Newtonian mechanics. In effect, Newton did for science what formalization did, or tried to do, for mathematics some three centuries later. There is indeed a profound parallel between Newtonian particle mechanics and the pure syntax of formalizations; in each case, everything is supposed to be generated from structureless, meaningless elements (particles in one case, symbols in the other), pushed around according to definite rules (forces in one case, production rules in the other). In each case, all that ultimately matters is the spatial disposition of these elements, their *configuration*. The concept of state does for particle mechanics what proposition does for formalism; it expresses a "meaningful" configuration of basic elements on which the syntactic rules can act. The sequence of state transitions of a system of particles, governed by Newton's laws of motion, starting from some given initial configuration, is then the analog of a *theorem* in a formalism, generated from an initial proposition (hypothesis) under the influence of the production rules. And just as formalization in mathematics believed that everything could be formalized *without loss*, so that all truth could be recaptured in terms of syntax alone, so particle mechanics came to believe that every material behavior could be, and should be, and indeed must be, reduced to purely syntactical sequences of configurations in an underlying system of particles.

Hence the power of the belief in reductionism, the scientific equivalent of the formalist faith in syntax. Though of course Newtonian mechanics has had to be supplemented and generalized repeatedly, the basic faith in syntax has not changed; indeed, it has been bolstered and made more credible by these very improvements. And there has as yet been no Gödel in physics to challenge that credibility directly. But there *is* biology.

The syntactical ideas, first and most potently manifested in Newtonian particle mechanics, have had another ramification; namely, the *form* taken by mechanical description (i.e., a set of states, on which are superimposed a set of rules governing change of state) has become the universal currency for describing systems of any kind. Thus, although in most cases we may not be able to get our hands on the underlying particles, and the rules governing them, we still exclusively utilize the Newtonian *language;* whether the field be chemical kinetics, or population dynamics, or economics, or power engineering, we still try to define an appropriate set of states, and a set of dynamical laws, in terms of which the behaviors of interest of a system can be generated and understood. From a purely reductionistic viewpoint, such an approach is only a stopgap, a formalism not yet fully formalized, but while we are waiting for the ultimate reduction to be effected, we can still use the form of the reductionistic language, if not yet its substance. And in fact, this is exactly what we do. However, when we divorce the language of mechanics from its susbtance, we run into certain problems. In Newtonian mechanics we have the luxury of being very explicit about what *state* means and what its properties are. When we leave the substance of mechanics behind, the concept becomes murky, and it becomes increasingly easy to equivocate on it. Nevertheless, it remains basic to every way we presently have of dealing with material reality; indeed, it is generally taken for granted as the starting point for every mode of precise scientific investigation. Hence it is appropriate that we begin our own analysis here.